Essential MATLAB

for Engineers and Scientists

Essential MATLAB
for Engineers and Scientists
Seventh Edition

Brian D. Hahn

Daniel T. Valentine

ELSEVIER

AMSTERDAM • BOSTON • HEIDELBERG • LONDON
NEW YORK • OXFORD • PARIS • SAN DIEGO
SAN FRANCISCO • SINGAPORE • SYDNEY • TOKYO

Academic Press is an imprint of Elsevier

Academic Press is an imprint of Elsevier
50 Hampshire Street, 5th Floor, Cambridge, MA 02139, United States
525 B Street, Suite 1800, San Diego, CA 92101-4495, United States
The Boulevard, Langford Lane, Kidlington, Oxford OX5 1GB, United Kingdom
125, London Wall, EC2Y, 5AS, United Kingdom

Library of Congress Cataloging-in-Publication Data
A catalog record for this book is available from the Library of Congress.

British Library Cataloguing-in-Publication Data
A catalogue record for this book is available from the British Library.

ISBN: 978-0-08-102997-8

For information on all Academic Press publications
visit our website at https://www.elsevierdirect.com

 Working together
to grow libraries in
developing countries

www.elsevier.com • www.bookaid.org

Publisher: Katey Birtcher
Acquisition Editor: Steve Merken
Editorial Project Manager: Katerina Zaliva
Production Project Manager: Nilesh Kumar Shah
Designer: Matthew Limbert

Typeset by VTeX

Printed in Great Britain
Last digit is the print number: 10 9 8 7 6 5 4 3

Contents

PREFACE ... xv

Part 1 **Essentials**

CHAPTER 1 Introduction ...3

 1.1 Using MATLAB ..5
 1.1.1 Arithmetic ...6
 1.1.2 Variables ...7
 1.1.3 Mathematical functions8
 1.1.4 Functions and commands8
 1.1.5 Vectors ...9
 1.1.6 Linear equations11
 1.1.7 Tutorials and demos12
 1.2 The desktop ..13
 1.2.1 Using the Editor and running a script14
 1.2.2 Help, Publish and View16
 1.2.3 Symbolics in live scripts19
 1.2.4 APPS ...20
 1.2.5 Additional features22
 1.3 Sample program ..24
 1.3.1 Cut and paste ..24
 1.3.2 Saving a program: Script files26
 1.3.3 A program in action28
 Summary ...29
 Exercises ..30
 Appendix 1.A Supplementary material30

CHAPTER 2 MATLAB Fundamentals ...31

 2.1 Variables ...31
 2.1.1 Case sensitivity32
 2.2 The workspace ..32
 2.2.1 Adding commonly used constants to the workspace33

2.3 Arrays: Vectors and matrices 34
 2.3.1 Initializing vectors: Explicit lists 34
 2.3.2 Initializing vectors: The colon operator 35
 2.3.3 The `linspace` and `logspace` functions 36
 2.3.4 Transposing vectors 37
 2.3.5 Subscripts .. 37
 2.3.6 Matrices .. 38
 2.3.7 Capturing output .. 39
 2.3.8 Structure plan ... 39
2.4 Vertical motion under gravity 40
2.5 Operators, expressions, and statements 42
 2.5.1 Numbers .. 43
 2.5.2 Data types ... 43
 2.5.3 Arithmetic operators 44
 2.5.4 Operator precedence 44
 2.5.5 The colon operator .. 45
 2.5.6 The transpose operator 46
 2.5.7 Arithmetic operations on arrays 46
 2.5.8 Expressions ... 47
 2.5.9 Statements .. 48
 2.5.10 Statements, commands, and functions 49
 2.5.11 Formula vectorization 49
2.6 Output ... 52
 2.6.1 The `disp` statement 52
 2.6.2 The `format` command 54
 2.6.3 Scale factors .. 55
2.7 Repeating with `for` ... 55
 2.7.1 Square roots with Newton's method 56
 2.7.2 Factorials! ... 57
 2.7.3 Limit of a sequence 58
 2.7.4 The basic `for` construct 58
 2.7.5 `for` in a single line 60
 2.7.6 More general `for` .. 60
 2.7.7 Avoid `for` loops by vectorizing! 60
2.8 Decisions ... 63
 2.8.1 The one-line `if` statement 63
 2.8.2 The `if-else` construct 65
 2.8.3 The one-line `if-else` statement 66
 2.8.4 `elseif` ... 66
 2.8.5 Logical operators ... 68
 2.8.6 Multiple `if`s versus `elseif` 68
 2.8.7 Nested `if`s ... 70
 2.8.8 Vectorizing `if`s? .. 71
 2.8.9 The `switch` statement 71
2.9 Complex numbers .. 71

	Summary	73
	Exercises	75
	Appendix 2.A Supplementary material	81
CHAPTER 3	**Program Design and Algorithm Development**	83
	3.1 The program design process	84
	3.1.1 The projectile problem	87
	3.2 Programming MATLAB functions	92
	3.2.1 Inline objects: Harmonic oscillators	92
	3.2.2 MATLAB function: $y = f(x)$	94
	Summary	96
	Exercises	97
CHAPTER 4	**MATLAB Functions and Data Import-Export Utilities**	99
	4.1 Common functions	99
	4.2 Importing and exporting data	104
	4.2.1 The `load` and `save` commands	104
	4.2.2 Exporting text (ASCII) data	104
	4.2.3 Importing text (ASCII) data	105
	4.2.4 Exporting binary data	105
	4.2.5 Importing binary data	106
	Summary	106
	Exercises	106
CHAPTER 5	**Logical Vectors**	109
	5.1 Examples	110
	5.1.1 Discontinuous graphs	110
	5.1.2 Avoiding division by zero	111
	5.1.3 Avoiding infinity	112
	5.1.4 Counting random numbers	113
	5.1.5 Rolling dice	114
	5.2 Logical operators	115
	5.2.1 Operator precedence	116
	5.2.2 Danger	116
	5.2.3 Logical operators and vectors	117
	5.3 Subscripting with logical vectors	118
	5.4 Logical functions	119
	5.4.1 Using `any` and `all`	120
	5.5 Logical vectors instead of `elseif` ladders	121
	Summary	123
	Exercises	124
	Appendix 5.A Supplementary material	126
CHAPTER 6	**Matrices and Arrays**	127
	6.1 Matrices	127
	6.1.1 A concrete example	127
	6.1.2 Creating matrices	129

	6.1.3	Subscripts	129
	6.1.4	Transpose	130
	6.1.5	The colon operator	130
	6.1.6	Duplicating rows and columns: Tiling	133
	6.1.7	Deleting rows and columns	134
	6.1.8	Elementary matrices	135
	6.1.9	Specialized matrices	136
	6.1.10	Using MATLAB functions with matrices	137
	6.1.11	Manipulating matrices	138
	6.1.12	Array (element-by-element) operations on matrices	138
	6.1.13	Matrices and `for`	139
	6.1.14	Visualization of matrices	139
	6.1.15	Vectorizing nested `for`s: loan repayment tables	139
	6.1.16	Multi-dimensional arrays	142
6.2	Matrix operations		143
	6.2.1	Matrix multiplication	143
	6.2.2	Matrix exponentiation	145
6.3	Other matrix functions		145
6.4	Population growth: Leslie matrices		146
6.5	Markov processes		149
	6.5.1	A random walk	150
6.6	Linear equations		152
	6.6.1	MATLAB's solution	153
	6.6.2	The residual	153
	6.6.3	Over-determined systems	154
	6.6.4	Under-determined systems	155
	6.6.5	Ill conditioning	155
	6.6.6	Matrix division	156
6.7	Sparse matrices		157
	Summary		160
	Exercises		160
CHAPTER 7	Function M-files		163
7.1	Example: Newton's method again		163
7.2	Basic rules		165
	7.2.1	Subfunctions	170
	7.2.2	Private functions	170
	7.2.3	P-code files	170
	7.2.4	Improving M-file performance with the Profiler	171
7.3	Function handles		171
7.4	Command/function duality		173
7.5	Function name resolution		173

7.6 Debugging M-files...174
 7.6.1 Debugging a script...............................174
 7.6.2 Debugging a function............................176
7.7 Recursion ..176
 Summary ...177
 Exercises ..179
 Appendix 7.A Supplementary material180

CHAPTER 8 Loops ..181
8.1 Determinate repetition with `for`.....................181
 8.1.1 Binomial coefficient............................181
 8.1.2 Update processes..............................182
 8.1.3 Nested `for`s184
8.2 Indeterminate repetition with `while`184
 8.2.1 A guessing game..............................184
 8.2.2 The `while` statement........................185
 8.2.3 Doubling time of an investment.............186
 8.2.4 Prime numbers................................187
 8.2.5 Projectile trajectory...........................188
 8.2.6 `break` and `continue`.........................190
 8.2.7 Menus ...191
 Summary ..192
 Exercises ..192

CHAPTER 9 MATLAB Graphics.......................................197
9.1 Basic 2-D graphs197
 9.1.1 Labels ...198
 9.1.2 Multiple plots on the same axes199
 9.1.3 Line styles, markers and color200
 9.1.4 Axis limits......................................200
 9.1.5 Multiple plots in a figure: `subplot`...........202
 9.1.6 `figure`, `clf` and `cla`203
 9.1.7 Graphical input................................203
 9.1.8 Logarithmic plots..............................203
 9.1.9 Polar plots204
 9.1.10 Plotting rapidly changing mathematical
 functions: `fplot`.............................205
 9.1.11 The Property Editor...........................206
9.2 3-D plots ...206
 9.2.1 `plot3` ...206
 9.2.2 Animated 3-D plots with `comet3`............207
 9.2.3 Mesh surfaces................................207
 9.2.4 Contour plots.................................209
 9.2.5 Cropping a surface with `NaNs`211
 9.2.6 Visualizing vector fields......................211

9.2.7	Visualization of matrices	212
9.2.8	Rotation of 3-D graphs	213
9.3	Handle Graphics	214
9.3.1	Getting handles	215
9.3.2	Graphics object properties and how to change them	216
9.3.3	A vector of handles	218
9.3.4	Graphics object creation functions	218
9.3.5	Parenting	219
9.3.6	Positioning figures	219
9.4	Editing plots	220
9.4.1	Plot edit mode	220
9.4.2	Property Editor	221
9.5	Animation	222
9.5.1	Animation with Handle Graphics	223
9.6	Color etc.	225
9.6.1	Colormaps	225
9.6.2	Color of surface plots	227
9.6.3	Truecolor	228
9.7	Lighting and camera	229
9.8	Saving, printing and exporting graphs	230
9.8.1	Saving and opening figure files	230
9.8.2	Printing a graph	230
9.8.3	Exporting a graph	230
	Summary	231
	Exercises	232
CHAPTER 10	**Vectors as Arrays and Other Data Structures**	237
10.1	Update processes	237
10.1.1	Unit time steps	238
10.1.2	Non-unit time steps	240
10.1.3	Using a function	242
10.1.4	Exact solution	243
10.2	Frequencies, bar charts and histograms	244
10.2.1	A random walk	244
10.2.2	Histograms	245
10.3	Sorting	246
10.3.1	Bubble Sort	246
10.3.2	MATLAB's sort	248
10.4	Structures	249
10.5	Cell arrays	251
10.5.1	Assigning data to cell arrays	251
10.5.2	Accessing data in cell arrays	253
10.5.3	Using cell arrays	253
10.5.4	Displaying and visualizing cell arrays	254

10.6 Classes and objects ...254
Summary ..255

CHAPTER 11 Errors and Pitfalls ...257
11.1 Syntax errors..257
11.1.1 Incompatible vector sizes...........................257
11.1.2 Name hiding ..258
11.2 Logic errors...258
11.3 Rounding error ...259
Summary ..260
Chapter exercises ...260

Part 2 Applications

CHAPTER 12 Dynamical Systems ..265
12.1 Cantilever beam ...267
12.2 Electric current ..268
12.3 Free fall...271
12.4 Projectile with friction ...280
Summary ..284
Exercises ...284

CHAPTER 13 Simulation..285
13.1 Random number generation285
13.1.1 Seeding `rand` ...286
13.2 Spinning coins ...286
13.3 Rolling dice...287
13.4 Bacteria division..288
13.5 A random walk ...288
13.6 Traffic flow..290
13.7 Normal (Gaussian) random numbers...................293
Summary ..293
Exercises ...294

CHAPTER 14 Introduction to Numerical Methods299
14.1 Equations..299
14.1.1 Newton's method ..299
14.1.2 The Bisection method...................................301
14.1.3 `fzero` ..303
14.1.4 `roots` ..303
14.2 Integration ..304
14.2.1 The Trapezoidal rule304
14.2.2 Simpson's rule ..305
14.2.3 `quad` ..306
14.3 Numerical differentiation306
14.3.1 `diff` ..307

14.4 First-order differential equations308
 14.4.1 Euler's method ...308
 14.4.2 Example: Bacteria growth309
 14.4.3 Alternative subscript notation......................311
 14.4.4 A predictor-corrector method312
14.5 Linear ordinary differential equations (LODEs)313
14.6 Runge-Kutta methods ...313
 14.6.1 A single differential equation313
 14.6.2 Systems of differential equations: Chaos.....314
 14.6.3 Passing additional parameters to an ODE
 solver ...317
14.7 A partial differential equation318
 14.7.1 Heat conduction ..318
14.8 Complex variables and conformal mapping.............322
 Joukowski airfoil ..322
14.9 Other numerical methods323
 Summary ...325
 Exercises ..325

CHAPTER 15 Signal Processing ...329
15.1 Harmonic analysis ..330
15.2 Fast Fourier Transform (FFT)...................................335

CHAPTER 16 SIMULINK® Toolbox ...341
16.1 Mass-spring-damper dynamic system347
16.2 Bouncing ball dynamic system349
16.3 The van der Pol oscillator..352
16.4 The Duffing oscillator ...353
 Exercises ..355

CHAPTER 17 Symbolics Toolbox ..359
17.1 Algebra ...360
 17.1.1 Polynomials...361
 17.1.2 Vectors..363
 17.1.3 Matrices..364
17.2 Calculus..368
17.3 Laplace and Z transforms370
17.4 Generalized functions* ...371
17.5 Differential equations ...373
17.6 Implementation of funtool, MuPAD and help...........374
 17.6.1 The funtool ...374
 17.6.2 The MuPAD notebook* and Symbolic help...375
 Exercises ..377

APPENDIX A Syntax: Quick Reference ...379
A.1 Expressions ...379
A.2 Function M-files ...379

A.3 Graphics...379

A.4 `if` and `switch`...380

A.5 `for` and `while`..381

A.6 Input/output..381

A.7 `load`/`save`..382

A.8 Vectors and matrices..382

APPENDIX B Operators ...385

APPENDIX C Command and Function: Quick Reference387

C.1 General-purpose commands387

 C.1.1 Managing variables and the workspace387

 C.1.2 Files and the operating system....................387

 C.1.3 Controlling the Command Window..............388

 C.1.4 Starting and quitting MATLAB.....................388

C.2 Logical functions...388

C.3 MATLAB programming tools................................388

 C.3.1 Interactive input ...389

C.4 Matrices...389

 C.4.1 Special variables and constants389

 C.4.2 Time and date..389

 C.4.3 Matrix manipulation....................................389

 C.4.4 Specialized matrices....................................390

C.5 Mathematical functions.......................................390

C.6 Matrix functions ..391

C.7 Data analysis ..391

C.8 Polynomial functions ...392

C.9 Function functions ...392

C.10 Sparse matrix functions392

C.11 Character string functions392

C.12 File I/O functions...392

C.13 2D graphics ..393

C.14 3D graphics ..393

C.15 General ..393

APPENDIX D Solutions to Selected Exercises............................395

INDEX ..407

Preface

The main reason for a seventh edition of *Essential MATLAB for Engineers and Scientists* is to keep up with MATLAB, now in its latest version (9.5 Version R2018b). Like the previous editions, this one presents MATLAB as a problem-solving tool for professionals in science and engineering, as well as students in those fields, who have no prior knowledge of computer programming.

In keeping with the late Brian D. Hahn's objectives in previous editions, the seventh edition adopts an informal, tutorial style for its "teach-yourself" approach, which invites readers to experiment with MATLAB as a way of discovering how it works. It assumes that readers have never used this tool in their technical problem solving.

MATLAB, which stands for "Matrix Laboratory," is based on the concept of the matrix. Because readers may be unfamiliar with matrices, ideas and constructs are developed gradually, as the context requires. The primary audience for *Essential MATLAB* is scientists and engineers, and for that reason certain examples require some first-year college math, particularly in Part 2. However, these examples are self-contained and can be skipped without detracting from the development of readers' programming skills.

MATLAB can be used in two distinct modes. One, in keeping the modern-age craving for instant gratification, offers immediate execution of statements (or groups of statements) in the Command Window. The other, for the more patient, offers conventional programming by means of script files. Both modes are put to good use here: The former encouraging cut and paste to take full advantage of Windows' interactive environment. The latter stressing programming principles and algorithm development through structure plans.

Although most of MATLAB's basic ("essential") features are covered, this book is neither an exhaustive nor a systematic reference. This would not be in keeping with its informal style. For example, constructs such as for and if are not always treated, initially, in their general form, as is common in many texts, but are gradually introduced in discussions where they fit naturally. Even so, they are treated thoroughly here, unlike in other texts that deal with them only su-

perficially. For the curious, helpful syntax and function quick references can be found in the appendices.

Essential MATLAB is meant to be used in conjunction with the MATLAB software. The reader is expected to have the software at hand in order to work through the exercises and thus discover how MATLAB does what it is commanded to do. Learning any tool is possible only through hands-on experience. This is particularly true with computing tools, which produce correct answers only when the commands they are given and the accompanying data input are correct and accurate.

ACKNOWLEDGMENTS

I would like to thank Mary, Clara, Zoe Rae and Zach T. for their support and encouragement. I dedicate the seventh edition of *Essential MATLAB for Engineers and Scientists* to them.

Daniel T. Valentine

1

Essentials

Part 1 concerns those aspects of MATLAB that you need to know in order to come to grips with MATLAB's essentials and those of technical computing. Because this book is a tutorial, you are encouraged to use MATLAB extensively while you go through the text.

Introduction

THE OBJECTIVES OF THIS CHAPTER ARE:

- To enable you to use some simple MATLAB commands from the Command Window.
- To examine various MATLAB desktop and editing features.
- To learn some of the new features of the MATLAB R2018b Desktop.
- To learn to write scripts in the Editor and Run them from the Editor.
- To learn some of the new features associated with the tabs (in particular, the PUBLISH and APPS features).

CONTENTS

Using MATLAB . 5
Arithmetic 6
Variables 7
Mathematical
functions 8
Functions and
commands 8
Vectors 9
Linear equations.... 11
Tutorials and demos 12

The desktop 13
Using the Editor and
running a script 14
Help, Publish and
View 16
Symbolics in live
scripts 19
APPS 20
Additional features . 22

Sample program 24
Cut and paste 24
Saving a program:
Script files 26
Current directory 27
Running a script from
the Current Folder
browser 28
A program in action 28

Summary 29

Exercises 30

Supplementary
material 30

MATLAB is a powerful technical computing system for handling scientific and engineering calculations. The name MATLAB stands for Matrix Laboratory, because the system was designed to make matrix computations particularly easy. A matrix is an array of numbers organized in m rows and n columns. An example is the following $m \times n = 2 \times 3$ array:

$$A = \begin{pmatrix} 1 & 3 & 5 \\ 2 & 4 & 6 \end{pmatrix}$$

Any one of the elements in a matrix can be accessed by using the row and column indices that identify its location. The elements in this example are accessed as follows: $A(1, 1) = 1$, $A(1, 2) = 3$, $A(1, 3) = 5$, $A(2, 1) = 2$, $A(2, 2) = 4$, $A(2, 3) = 6$. The first index identifies the row number counted from top to bottom; the second index is the column number counted from left to right. This is the convention used in MATLAB to locate information in a matrix. A computer is useful because it can do numerous computations quickly, so operating on large numerical data sets listed in tables or matrices (or arrays) of rows and columns is quite efficient.

This book assumes that you are an engineer, a scientist or an undergraduate student majoring in a STEM (science, technology, engineering and mathematics) field. Hence, it is assumed that students and practitioners in STEM have been introduced to matrices in their mathematics courses prior to entering a college or university. On the other hand this book assumes that you have never used MATLAB to solve engineering or scientific problems and are interested in

3

discovering the technical computing capabilities of this tool. Hence, it is also assumed that you can find your way around the keyboard and know your operating system (e.g., Windows, UNIX or MAC-OS). Some experience using a computer is expected, e.g., doing word processing, doing basic text editing, etc.

One of the many things you will like about MATLAB (and that distinguishes it from many other computer programming systems, such as C++ and Java) is that you can use it *interactively*. This means you type some commands at the special MATLAB prompt and get results immediately. The problems solved in this way can be very simple, like finding a square root, or very complicated, like finding the solution to a system of differential equations. For many technical problems, you enter only one or two commands—MATLAB does most of the work for you.

There are three essential requirements for successful MATLAB applications:

- You must learn the *exact* rules for writing MATLAB statements and using MATLAB utilities.
- You must know the mathematics associated with the problem you want to solve.
- You must develop a logical plan of attack—the algorithm—for solving a particular problem.

This chapter is devoted mainly to the first requirement: learning some basic MATLAB rules. Computer programming is the process of writing a collection of instructions that performs a specific task when executed by a computer. In this book you are introduced to programming by using some of the capabilities of MATLAB to do technical computing.

With experience, you will be able to design, develop and implement computational and graphical tools to do relatively complex science and engineering problems. You will be able to adjust the look of MATLAB, modify the way you interact with it, and develop a toolbox of your own that helps you solve problems of interest. In other words, you can, with significant experience, customize your MATLAB working environment.

In the rest of this chapter we will look at some simple examples. Don't be concerned about understanding exactly what is happening. Understanding will come with the work you need to do in later chapters. It is very important for you to practice with MATLAB to learn how it works. Once you have grasped the basic rules in this chapter, you will be prepared to master many of those presented in the next chapter and in the Help files provided with MATLAB. This will help you go on to solve more interesting and substantial problems. In the last section of this chapter you will take a quick tour of the MATLAB desktop.

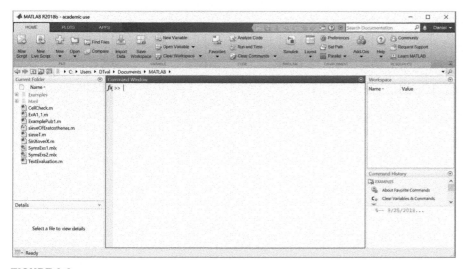

FIGURE 1.1
MATLAB desktop illustrating the Home task bar (version 2018b).

1.1 USING MATLAB

Either MATLAB must be installed on your computer or you must have access to a network where it is available. Throughout this book the latest version at the time of writing is assumed (version R2018b).

To start from Windows, double-click the MATLAB icon on your Windows desktop. To start from UNIX, type matlab at the operating system prompt. To start from MAC-OS open X11 (i.e., open an X-terminal window), then type matlab at the prompt. The MATLAB desktop opens as shown in Fig. 1.1. The window in the desktop that concerns us for now is the Command Window, where the special >> prompt appears. This prompt means that MATLAB is waiting for a command. You can quit at any time with one of the following ways:

- Click the **X** (close box) in the upper right-hand corner of the MATLAB desktop.
- Type quit or exit at the Command Window prompt followed by pressing the 'enter' key.

Starting MATLAB automatically creates a folder named MATLAB in the user's Documents Folder. This feature is quite convenient because it is the default working folder. It is in this folder that anything saved from the Command Window will be saved. Now you can experiment with MATLAB in the Command Window. If necessary, make the Command Window active by placing the cursor in the Command Window and left-clicking the mouse button anywhere inside its border.

1.1.1 Arithmetic

Since we have experience doing arithmetic, we want to examine if MATLAB does it correctly. This is a required step to gain confidence in any tool and in our ability to use it.

Type 2+3 after the >> prompt, followed by **Enter** (press the **Enter** key) as indicated by <**Enter**>:

>> 2+3 <**Enter**>

Commands are only carried out when you enter them. The answer in this case is, of course, 5. Next try

>> 3-2 <**Enter**>
>> 2*3 <**Enter**>
>> 1/2 <**Enter**>
>> 2^3 <**Enter**>
>> 2\11 <**Enter**>

What about (1)/(2) and (2)^(3)? Can you figure out what the symbols *, /, and ^ mean? Yes, they are multiplication, division and exponentiation. The backslash means the denominator is to the left of the symbol and the numerator is to the right; the result for the last command is 5.5. This operation is equivalent to 11/2.

Now enter the following commands:

>> 2 .* 3 <**Enter**>
>> 1 ./ 2 <**Enter**>
>> 2 .^ 3 <**Enter**>

A period in front of the *, /, and ^, respectively, does not change the results because the multiplication, division, and exponentiation is done with single numbers. (An explanation for the need for these symbols is provided later when we deal with arrays of numbers.)

Here are hints on creating and editing command lines:

- The line with the >> prompt is called the *command line*.
- You can edit a MATLAB command before pressing **Enter** by using various combinations of the **Backspace**, **Left-arrow**, **Right-arrow**, and **Del** keys. This helpful feature is called *command-line editing*.
- You can select (and edit) commands you have entered using **Up-arrow** and **Down-arrow**. Remember to press **Enter** to have the command carried out (i.e., to *run* or to *execute* the command).

- MATLAB has a useful editing feature called *smart recall*. Just type the first few characters of the command you want to recall. For example, type the characters 2* and press the **Up-arrow** key—this recalls the most recent command starting with 2*.

How do you think MATLAB would handle 0/1 and 1/0? Try it. If you insist on using ∞ in a calculation, which you may legitimately wish to do, type the symbol Inf (short for *infinity*). Try 13+Inf and 29/Inf.

Another special value that you may meet is NaN, which stands for *Not-a-Number*. It is the answer to calculations like 0/0.

1.1.2 Variables

Now we will assign values to variables to do arithmetic operations with the variables. First enter the command (*statement* in programming jargon) a = 2. The MATLAB command line should look like this:

>> a = 2 <**Enter**>

The a is a *variable*. This statement *assigns* the value of 2 to a. (Note that this value is displayed immediately after the statement is executed.) Now try entering the statement a = a + 7 followed on a new line by a = a * 10. Do you agree with the final value of a? Do we agree that it is 90?

Now enter the statement

>> b = 3; <**Enter**>

The semicolon (;) prevents the value of b from being displayed. However, b still has the value 3, as you can see by entering without a semicolon:

>> b <**Enter**>

Assign any values you like to two variables x and y. Now see if you can assign the *sum* of x and y to a third variable z in a single statement. One way of doing this is

>> x = 2; y = 3; <**Enter**>
>> z = x + y <**Enter**>

Notice that, in addition to doing the arithmetic with variables with assigned values, several commands separated by semicolons (or commas) can be put on one line.

1.1.3 Mathematical functions

MATLAB has all of the usual mathematical functions found on a scientific-electronic calculator, like `sin`, `cos`, and `log` (meaning the *natural* logarithm). See Appendix C for many more examples.

- Find $\sqrt{\pi}$ with the command `sqrt(pi)`. The answer should be 1.7725. Note that MATLAB knows the value of `pi` because it is one of its many built-in functions.
- Trigonometric functions like `sin(x)` expect the argument `x` to be in *radians*. Multiply degrees by $\pi/180$ to get radians. For example, use MATLAB to calculate sin(90°). The answer should be 1 (`sin(90*pi/180)`).
- The exponential function e^x is computed in MATLAB as `exp(x)`. Use this information to find e and $1/e$ (2.7183 and 0.3679).

Because of the numerous built-in functions like `pi` or `sin`, care must be taken in the naming of user-defined variables. Names should not duplicate those of built-in functions without good reason. This problem can be illustrated as follows:

```
>> pi = 4 <Enter>
>> sqrt(pi) <Enter>
>> whos <Enter>
>> clear pi <Enter>
>> whos <Enter>
>> sqrt(pi) <Enter>
>> clear <Enter>
>> whos <Enter>
```

Note that `clear` executed by itself clears all local variables in the *workspace* (the workspace is where the local variables defined on command lines are stored; see the Workspace pane on the right side of the default desktop); >> `clear pi` clears the locally defined variable `pi`. In other words, if you decide to redefine a built-in function or command, the new value is used! The command `whos` is executed to determine the list of local variables or commands presently in the workspace. The first execution of the command `pi = 4` in the above example displays your redefinition of the built-in `pi`: a 1-by-1 (or `1x1`) *double array*, which means this *data type* was created when `pi` was assigned a number.

1.1.4 Functions and commands

MATLAB has numerous general functions. Try `date` and `calendar` for starters. It also has numerous *commands*, such as `clc` (for *clear command window*). `help` is one you will use a lot (see below). The difference between functions and commands is that functions usually return with a value (e.g., the date), while

commands tend to change the environment in some way (e.g., clearing the screen or saving some statements to the workspace).

1.1.5 Vectors

Variables such as a and b that were used in Section 1.1.2 above are called *scalars*; they are single-valued. MATLAB also handles *vectors* (generally referred to as *arrays*), which are the key to many of its powerful features. The easiest way of defining a vector where the elements (components) increase by the same amount is with a statement like

>> x = 0 : 10; <**Enter**>

The symbol between the 0 and the 10 is a colon (:). There is no need to leave a space on either side of it, except to make it more readable. Enter x to check that x is a vector; it is a *row vector*—consisting of 1 row and 11 columns. Type the following command to verify that this is the case:

>> size(x) <**Enter**>

Part of the real power of MATLAB is illustrated by the fact that other vectors can now be defined (or created) in terms of the just defined vector x. Try

>> y = 2 .* x <**Enter**>
>> w = y ./ x <**Enter**>

and

>> z = sin(x) <**Enter**>

(no semicolons). Note that the first command line creates a vector y by multiplying each element of x by the factor 2. The second command line is an array operation, creating a vector w by taking each element of y and dividing it by the corresponding element of x. Since each element of y is two times the corresponding element of x, the vector w is a row vector of 11 elements all equal to 2. Finally, z is a vector with sin(x) as its elements.

To draw a reasonably nice graph of $\sin(x)$, simply enter the following commands:

>> x = 0 : 0.1 : 10; <**Enter**>
>> y = sin(x); <**Enter**>
>> plot(x,y), grid <**Enter**>

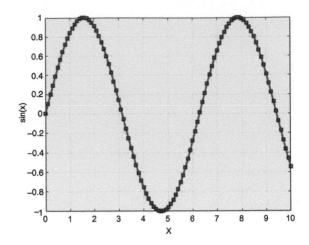

FIGURE 1.2
Figure window.

The graph appears in a separate figure window. To draw the graph of the sine function illustrated in Fig. 1.2 replace the last line above with

>> plot(x,y,'-rs','LineWidth',2,'MarkerEdgeColor','k','MarkerSize',5),grid
 <Enter>
>> xlabel(' x '), ylabel(' sin(x) ') **<Enter>**
>> whitebg('y') **<Enter>**

You can select the Command Window or figure windows by clicking anywhere inside them. The **Windows** pull-down menus can be used in any of them.

Note that the first command line above has three numbers after the equal sign. When three numbers are separated by two colons in this way, the middle number is the *increment*. The increment of 0.1 was selected to give a reasonably smooth graph. The command grid following the comma in the last command line adds a grid to the graph.

Modifying the plot function as illustrated above, of the many options available within this function, four were selected. A comma was added after the variable y followed by '-rs'. This selects a solid red line (-r) to connect the points at which the sine is computed; they are surrounded by square (s) markers in the figure. The line width is increased to 2 and the marker edge color is black (k) with size 5. Axis labels and the background color were changed with the statements following the plot command. (Additional changes in background color, object colors, etc., can be made with the figure properties editor; it can be found in

the pull-down menu under **Edit** in the figure toolbar. Many of the colors in the figures in this book were modified with the figure-editing tools.)

If you want to see more cycles of the sine graph, use command-line editing to change `sin(x)` to `sin(2*x)`.

Try drawing the graph of `tan(x)` over the same domain. You may find aspects of your graph surprising. To help examine this function you can improve the graph by using the command `axis([0 10 -10 10])` as follows:

```
>> x = 1:0.1:10; <Enter>
>> z = tan(x); <Enter>
>> plot(x,z),axis([0 10 -10 10]) <Enter>
```

An alternative way to examine mathematical functions graphically is to use the following command:

```
>> ezplot('tan(x)') <Enter>
```

The apostrophes around the function `tan(x)` are important in the `ezplot` command. Note that the default domain of x in `ezplot` is *not* 0 to 10.

A useful Command Window editing feature is *tab completion*: Type the first few letters of a MATLAB name and then press **Tab**. If the name is unique, it is automatically completed. If it is not unique, press **Tab** a second time to see all the possibilities. Try by typing `ta` at the command line followed by **Tab** twice.

1.1.6 Linear equations

Systems of linear equations are very important in engineering and scientific analysis. A simple example is finding the solution to two simultaneous equations:

$$x + 2y = 4$$
$$2x - y = 3$$

Here are two approaches to the solution.

Matrix method. Type the following commands (exactly as they are):

```
>> a = [1 2; 2 -1]; <Enter >
>> b = [4; 3]; <Enter >
>> x = a\b <Enter >
```

The result is

```
x =

   2
   1
```

i.e., $x = 2$, $y = 1$.

Built-in `solve` *function.* Type the following commands (exactly as they are):

```
>> syms x y; [x,y] = solve(x+2*y-4, 2*x - y-3) <Enter >
>> whos <Enter >
>> x = double(x), y=double(y) <Enter >
>> whos <Enter >
```

The function *double* converts x and y from symbolic objects (another *data type* in MATLAB) to double arrays (i.e., the numerical-variable data type associated with an assigned number).

To check your results, after executing either approach, type the following commands (exactly as they are):

```
>> x + 2*y % should give ans = 4 <Enter >
>> 2*x - y % should give ans = 3 <Enter >
```

The % symbol is a *flag* that indicates all information to the right is *not* part of the command but a comment. (We will examine the need for comments when we learn to develop coded programs of command lines later on.)

1.1.7 Tutorials and demos

If you want a spectacular sample of what MATLAB has to offer, type the command `demo` on the command line. After entering this command the Help documentation is opened at MATLAB Examples (see Fig. 1.3). Left-click on "Getting Started". This points you to the list of tutorials and demonstrations of MATLAB applications that are at your disposal. Click on any of the other topics to learn more about the wealth of capabilities of MATLAB. You may wish to review the tutorials appropriate to the topics you are examining as part of your technical computing needs. Click on "View more MATLAB examples" and scroll down to "Animations" and to "Images" to learn more about the features of MATLAB to produce motion pictures as a way to analyze various unsteady problems. New features are continually added to MATLAB; all new features are reported on the MathWorks website. MathWorks is the company that sells and supports the continued development of enhancements for MATLAB and SIMULINK and a variety of toolboxes that they offer to the science, engineering, technology and mathematics communities.

FIGURE 1.3
The Help documentation on MATLAB Examples.

1.2 THE DESKTOP

A very useful feature of MATLAB R2018a is the fact that when you first open it, it creates the folder named MATLAB (if it does not already exist) in your Documents folder. The first time it does this, there are no items in the folder and, hence, the Current Folder panel will be empty. This new folder in your Documents is the default working folder where all the files you create are saved. The location of this folder is given in the first toolbar above the Command Window. The location is `C:\Users\Clara\Documents\MATLAB`. This format of the location was determined by pointing and left-clicking the mouse in the line just above the Command Window.

Let us examine the Desktop from the top down. On the left side of the top line you should see the name of the version of MATLAB running. In this case it is MATLAB R2018a. On the right side of the top line are three buttons. They are the underscore button, which allows you to minimize the size of the Desktop window, the rectangle button, which allows you to maximize the size of the Desktop, and the × button, which allows you to close MATLAB (see Fig. 1.4).

On the next line of the Desktop there are three tabs on the left side. The first tap is most forward in the figure and, hence, the Home toolbar is displayed (the tabs and the toolbars associated with the tabs are the main new features of this release of MATLAB). If you are already familiar with a previous release of MATLAB, you will find that these new features enhance significantly the use of MATLAB. In addition, all previously developed tools operate exactly as they

FIGURE 1.4
New Desktop Toolbar on MATLAB 2018a.

FIGURE 1.5
Editor opened in default location; it is in the center of the Desktop.

did in previous releases of MATLAB. The other two tabs are PLOTS and APPS. These features allow you to access tools within MATLAB by pointing and clicking and, hence, enhance the utilization of tools and toolboxes available within MATLAB. In addition, the APPS environment allows the user to create their own applications (or APPS).

1.2.1 Using the Editor and running a script

Point and click on the New Script icon on the left most side of the Home toolbar. Doing this opens the editor in the center of the Desktop as shown in Fig. 1.5. Note that three new tabs appear and that the tab that is visible is the Editor tab that is connected with the Editor. The other two tabs are Publish and View. The latter are useful when creating notebooks or other documents connected with your technical computing work. The application of these tools will be illustrated by an example later in this text.

Let us first consider using the Editor. Type into the Editor the following script:

```
% Example of one of the matrix inversion methods available in MATLAB
clear;clc
% Let us consider the following arbitrarily selected matrix:
A =magic(3)
% Let us evaluate its inverse as follows:
AI = inv(A)
% Let us check that it is an inverse:
IPredicted = A * AI
% This is the exact unitary matrix:
IM = eye(3)
% The is the difference between the exact and predict unitary
% matrix:
difference = IPredicted - IM
for m = 1:3
    for n = 1:3
        if difference(m,n) < eps;
            IPredicted(m,n) = IM(m,n);
        end
    end
end
IPredicted
IPredicted == IM
```

Then click on the Run button just under the tab named View. The first time the script is executed you are asked to name the file. The name used in this example is ExA1_1.m. If all lines are typed correctly (except the lines beginning with the symbol '%', because they are comments that have nothing to do with the sequence of commands in the script except that they help the reader understand what the script does), what shows up in the Command Window is as follows:

```
A =
     8     1     6
     3     5     7
     4     9     2
AI =
    0.1472   -0.1444    0.0639
   -0.0611    0.0222    0.1056
   -0.0194    0.1889   -0.1028
IPredicted =
    1.0000         0   -0.0000
   -0.0000    1.0000         0
    0.0000         0    1.0000
IM =
```

```
        1       0       0
        0       1       0
        0       0       1
difference =
    1.0e-15 *
        0               0      -0.1110
    -0.0278             0            0
     0.0694            0            0
IPredicted =
        1       0       0
        0       1       0
        0       0       1
ans =
        1       1       1
        1       1       1
        1       1       1
```

The `IPredicted` matrix is supposed to be the identity matrix, `IM`. The matrix `IPredicted` was determined by multiplying the matrix `A` by its numerically computed inverse, `AI`. The last print out of `IPredicted` is a modification of the original matrix; it was changed to the elements of the `IM` matrix if the difference between a predicted and an actual element of `IM` was less than `eps` = 2.2204e − 16. Since the result is identical to the identity matrix, this shows that the inverse was computed correctly (at least to within the computational error of the computing environment, i.e., 0 < `eps`). This conclusion is a result of the fact that the `ans` in the above example produced the logical result of 1 (or *true*) for all entries in the adjusted `IPredicted` matrix as logically compared with the corresponding entries in `IM`.

At this point in the exercise the Desktop looks like Fig. 1.6. The name of the file is `ExA1_1.m`. It appears in the Current Folder and it also appears in the Command History. Note that the Workspace is populated with the variables created by this script.

This concludes the introduction of the most important tools needed for most of the exercises in Essential MATLAB (i.e., in this text). In the next section we examine an example of some of the other new features of MATLAB R2018a.

1.2.2 Help, Publish and View

Publish is an easy way to create notebooks or other documents in `html` format. The conversion of the information typed into an M-file is published into a document that looks like the new **Help** environment. To open the help documents go to the top of the Desktop to the question mark. Left click on the question mark ?. The Help window opens up. Left click on the topic "MATLAB" to open up the window illustrated in Fig. 1.7. This also illustrates the new format of

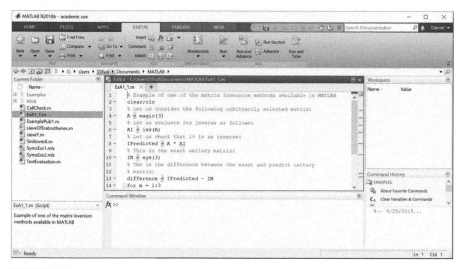

FIGURE 1.6
Sample script created and executed in the first example of this section.

FIGURE 1.7
Illustration of one of the pages in the online documentation for MATLAB.

the searchable documentation available within MATLAB R2018a. We want to compare this documentation with the kind you can PUBLISH yourself. To illustrate how easy it is to create documents like the MATLAB documents, let us consider the following simple example.

Click the New Script button to open up the editor (or type `edit` after the command prompt in the Command Window followed by tapping the enter key). The Editor tab is in the most forward position on the main taskbar.

Place the cursor on PUBLISH and left click on the mouse. This brings the PUBLISH toolbar forward. Left click Section with Title. Replace SECTION TITLE with PUBLISHING example. **Next, replace** DESCRIPTIVE TEXT **with**

```
%%
% This is an example to illustrate how easy it is to create a document
% in the PUBLISH environment.
%
% (1) This is an illustration of a formula created with a LaTeX command:
%
```

Next, click on Σ Inline LaTeX located in the "Insert Inline Markup" group. This leads to the addition of the equation $x^2+e^{\pi i}$. Following this equation add the text shown in the final script file shown below that ends with "clicked:". This is followed by a blank line and a command script; this command script is included to illustrate how MATLAB commands can be incorporated into published documents.

```
%%
% This is an example to illustrate how easy it is to create a document
% in the PUBLISH environment.
%
% (1) This is an illustration of a formula created with a LaTeX command:
%
%%
% $x^2+e^{\pi i}$
%
% (2) This is an illustration of how you can incorporate a MATLAB script
% in the document that is run when the Publish button below and to the
% right of View is clicked:

% Earth picture
load earth
image(X); colormap(map);
axis image
```

The final step is to left-click on Publish, which is just to the right and below View. The first window to appear is the one asking you to save the M-file. The name used in this example is ExamplePub1.m. After it appears in the Current Folder it is executed. A folder named html is automatically created and it contains the html document just created. The document is illustrated in Fig. 1.8.

Finally, the VIEW tab brings up a toolbar that allows you to change the configuration of the Editor window. From the authors point of view, the default Editor environment is fine as is especially for users who are beginning to use MATLAB for technical computing. Customizing your working environment is

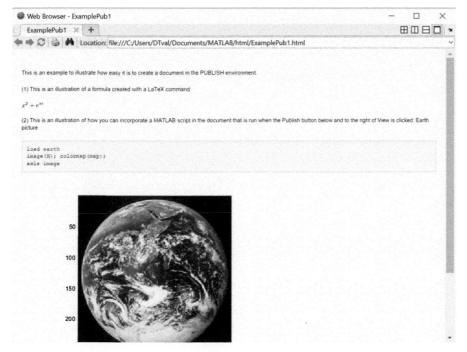

FIGURE 1.8
Sample document created in the Publish environment.

certainly possible in MATLAB. However, it is useful to learn how to deal with the default environment before deciding what needs to be changed to help satisfy your own requirements for using MATLAB.

1.2.3 Symbolics in live scripts

The Symbolic Math Toolbox is a useful tool to help you do symbolic mathematical analysis. It is convenient and the results are more presentable if we use this toolbox within a Live-Script environment. The portions of this type of file in which code is entered is essentially the same as what would be typed in the Work Space or in an ordinary script file. However, the display is much more presentable than the output displayed in the Work Space window of the Desktop. This will be evident by the examples reported in this section.

In the first set of examples we show how to differentiate a function and, subsequently, integrate a function using the Symbolic Tools. First let us open a Live-Editor file by clicking on the Live Editor icon below Home in the Home tool bar. We get the Live Editor in the center of the desktop as shown in Fig. 1.9. The gray box at the top of the file to be created is where the MATLAB scripts are typed into this editor.

FIGURE 1.9
The Live Editor window.

The simple examples in Fig. 1.10 illustrate the application of Symbolic Tools to do calculus. The examples in Fig. 1.11 illustrates the power of tools within MATLAB to study the shapes of functions that you may encounter in your work. One is the $\text{sech}^2(x)$ function which plays an important role in nonlinear wave theory. The second is the complete elliptic integral of the first kind, viz.,

$$K(m) = \int_0^{\pi/2} \frac{1}{\sqrt{1 - m^2 \sin^2 \theta}} \, d\theta$$

where $m = \pi/4$. This integral plays an important role in potential theory. This function was selected as an example because the second author encountered it in his work. The purpose of showing this is to illustrate how easy it is to explore some of the features of relatively complicated functions that arise in computational analysis of engineering problems. As an exercise, open up the help documents to find other functions, tutorials, and examples. The help documentation has improved tremendously over the years since MATLAB was first introduced.

1.2.4 APPS

Press the APPS tab to the left of the EDITOR tab to see the applications that came with your MATLAB suite of tools. There are a number of other APPS available from MathWorks. In addition, there is a capability for you to create your own APPS. Hence, if there is anything that we learn from our first experiences with MATLAB is that there is a lot to learn (a lifelong experience of learning) because of the wealth of technology incorporated in this technical computing environment. The fact that you can develop your own toolboxes,

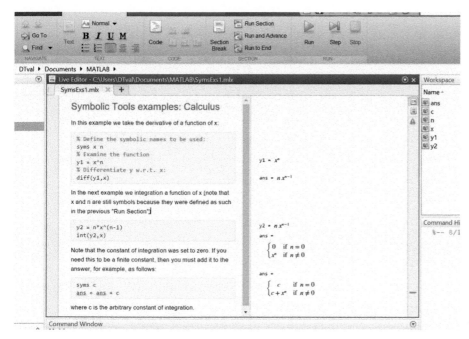

FIGURE 1.10
Simple examples of symbolic analysis of calculus problems.

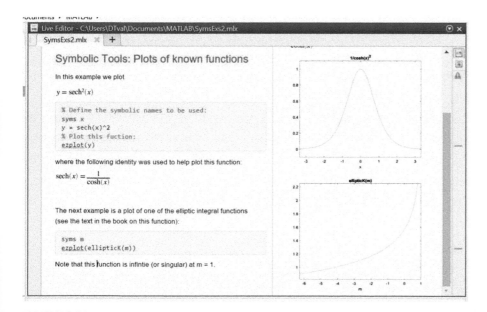

FIGURE 1.11
Simple examples of symbolic analysis of various important functions that arise in STEM fields.

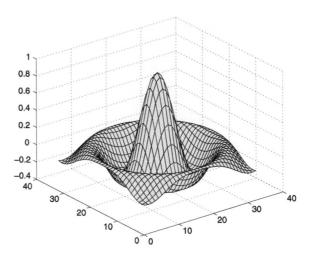

FIGURE 1.12
The Mexican hat.

your own APPS and you can customize your working environment (desktop arrangement, color backgrounds, fonts, graphical user interfaces and so on) provides real opportunities and useful experience in creating designs, creating useful tools and documenting your work.

1.2.5 Additional features

MATLAB has other good things. For example, you can generate a 10-by-10 (or 10×10) *magic square* by executing the command `magic(10)`, where the rows, columns, and main diagonal add up to the same value. Try it. In general, an $n \times n$ magic square has a row and column sum of $n(n^2 + 1)/2$.

You can even get a *contour* plot of the elements of a magic square. MATLAB pretends that the elements in the square are heights above sea level of points on a map, and draws the contour lines. `contour(magic(32))` looks interesting.

If you want to see the famous Mexican hat (Fig. 1.12), enter the following four lines (be careful not to make any typing errors):

```
>> [x y ] = meshgrid(-8 : 0.5 : 8); <Enter>
>> r = sqrt(x.^2 + y.^2) + eps; <Enter>
>> z = sin(r) ./ r; <Enter>
>> mesh(z); <Enter>
```

`surf(z)` generates a faceted (tiled) view of the surface. `surfc(z)` or `meshc(z)` draws a 2D contour plot under the surface. The command

```
>> surf(z), shading flat <Enter>
```

produces a nice picture by removing the grid lines.

The following animation is an extension of the Mexican hat graphic in Fig. 1.12. It uses a `for` loop that repeats the calculation from $n=-3$ to $n=3$ in increments of 0.05. It begins with a `for n=-3:0.05:3` command and ends with an `end` command and is one of the most important constructs in programming. The execution of the commands between the `for` and `end` statements repeat 121 times in this example. The `pause(0.05)` puts a time delay of 0.05 seconds in the `for` loop to slow the animation down, so the picture changes every 0.05 seconds until the end of the computation.

```
>> [x y]=meshgrid(-8:0.5:8); <Enter>
>> r=sqrt( x.^2+y.^2)+eps; <Enter>
>> for n=-3:0.05:3; <Enter>
>> z=sin(r.*n)./r; <Enter>
>> surf(z), view(-37, 38), axis([0,40,0,40,-4,4]); <Enter>
>> pause(0.05) <Enter>
>> end <Enter>
```

You can examine sound with MATLAB in any number of ways. One way is to listen to the signal. If your PC has a speaker, try

```
>> load handel <Enter>
>> sound(y,Fs) <Enter>
```

for a snatch of Handel's Hallelujah Chorus. For different sounds try loading `chirp`, `gong`, `laughter`, `splat`, and `train`. You have to run `sound(y,Fs)` for each one.

If you want to see a view of the Earth from space, try

```
>> load earth <Enter>
>> image(X); colormap(map) <Enter>
>> axis image <Enter>
```

To enter the matrix presented at the beginning of this chapter into MATLAB, use the following command:

```
>> A = [1 3 5; 2 4 6] <Enter>
```

On the next line after the command prompt, type `A(2,3)` to pluck the number from the second row, third column.

There are a few humorous functions in MATLAB. Try `why` (why not?) Then try `why(2)` twice. To see the MATLAB code that does this, type the following command:

>> `edit why` <**Enter**>

Once you have looked at this file, close it via the pull-down menu by clicking `File` at the top of the Editor desktop window and then `Close Editor`; do *not* save the file, in case you accidently typed something and modified it.

The `edit` command will be used soon to illustrate the creation of an *M-file* like `why.m` (the name of the file executed by the command `why`). You will create an M-file after we go over some of the basic features of the MATLAB desktop. More details on creating programs in the MATLAB environment will be explained in Chapter 2.

1.3 SAMPLE PROGRAM

In Section 1.1 we saw some simple examples of how to use MATLAB by entering single commands or statements at the MATLAB prompt. However, you might want to solve problems which MATLAB can't do in one line, like finding the roots of a quadratic equation (and taking all the special cases into account). A *collection* of statements to solve such a problem is called a *program*. In this section we look at the mechanics of writing and running two short programs, without bothering too much about how they work—explanations will follow in the next chapter.

1.3.1 Cut and paste

Suppose you want to draw the graph of $e^{-0.2x} \sin(x)$ over the domain 0 to 6π, as shown in Fig. 1.13. The Windows environment lends itself to nifty *cut and paste* editing, which you would do well to master. Proceed as follows.

From the MATLAB desktop select **File** -> **New** -> **Script**, or click the new file button on the desktop toolbar (you could also type `edit` in the Command Window followed by Enter). This action opens an Untitled window in the Editor/Debugger. You can regard this for the time being as a 'scratch pad' in which to write programs. Now type the following two lines in the Editor, exactly as they appear here:

```
x = 0 : pi/20 : 6 * pi;
plot(x, exp(-0.2*x) .* sin(x), 'k'),grid
```

Incidentally, that *is* a dot (full stop, period) in front of the second $*$ in the second line—a more detailed explanation later! The additional argument `'k'`

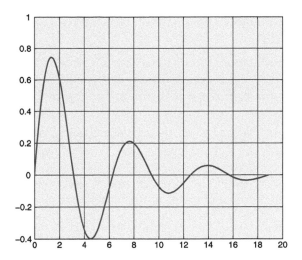

FIGURE 1.13
$e^{-0.2x}\sin(x)$.

for `plot` will draw a black graph, just to be different. Change `'k'` to `'r'` to generate a red graph if you prefer.

Next, move the mouse pointer (which now looks like a very thin capital I) to the left of the x in the first line. Keep the left mouse button down while moving the mouse pointer to the end of the second line. This process is called *dragging*. Both lines should be highlighted at this stage, probably in blue, to indicate that they have been *selected*.

Select the **Edit** menu in the Editor window, and click on **Copy** (or just use the keyboard shortcut **Ctrl+C**). This action copies the highlighted text to the Windows *clipboard*, assuming that your operating system is Windows.

Now go back to the Command Window. Make sure the cursor is positioned at the ≫ prompt (click there if necessary). Select the **Edit** menu, and click on **Paste** (or use the **Ctrl+V** shortcut). The contents of the clipboard will be copied into the Command Window. To execute the two lines in the program, press **Enter**. The graph should appear in a figure window.

This process, from highlighting (selecting) text in the Editor, to copying it into the Command Window, is called 'cut and paste' (more correctly 'copy and paste' here, since the original text is copied from the Editor, rather than being cut from it). It's well worth practicing until you have it right.

If you need to correct the program, go back to the Editor, click at the position of the error (this moves the *insertion point* to the right place), make the correction, and cut and paste again. Alternatively, you can use command-line editing to correct mistakes. As yet another alternative, you can paste from the Command History window (which incidentally goes back over many previous

sessions). To select multiple lines in the Command History window keep **Ctrl** down while you click.

If you prefer, you can enter multiple lines directly in the Command Window. To prevent the whole group from running until you have entered the last line use **Shift+Enter** after each line until the last. Then press **Enter** to run all the lines.

As another example, suppose you have $1000 saved in the bank. Interest is compounded at the rate of 9 per cent per year. What will your bank balance be after one year? Now, if you want to write a MATLAB program to find your new balance, you must be able to do the problem yourself in principle. Even with a relatively simple problem like this, it often helps first to write down a rough *structure plan*:

1. Get the data (initial balance and interest rate) into MATLAB.
2. Calculate the interest (9 per cent of $1000, i.e., $90).
3. Add the interest to the balance ($90 + $1000, i.e., $1090).
4. Display the new balance.

Go back to the Editor. To clear out any previous text, select it as usual by dragging (or use **Ctrl+A**), and press the **Del** key. By the way, to de-select highlighted text, click anywhere outside the selection area. Enter the following program, and then cut and paste it to the Command Window.

```
balance = 1000;
rate = 0.09;
interest = rate * balance;
balance = balance + interest;
disp( 'New balance:' );
disp( balance );
```

When you press **Enter** to run it, you should get the following output in the Command Window:

```
New balance:
        1090
```

1.3.2 Saving a program: Script files

We have seen how to cut and paste between the Editor and the Command Window in order to write and run MATLAB programs. Obviously you need to save the program if you want to use it again later.

To save the contents of the Editor, select **File** -> **Save** from the Editor menubar. A **Save file as:** dialogue box appears. Select a folder and enter a file name,

which *must* have the extension `.m`, in the **File name:** box, e.g., `junk.m`. Click on **Save**. The Editor window now has the title `junk.m`. If you make subsequent changes to `junk.m` in the Editor, an asterisk appears next to its name at the top of the Editor until you save the changes.

A MATLAB program saved from the Editor (or any ASCII text editor) with the extension `.m` is called a *script file*, or simply a *script*. (MATLAB *function* files also have the extension `.m`. We therefore refer to both script and function files generally as *M-files*.) The special significance of a script file is that, if you enter its name at the command-line prompt, MATLAB carries out each statement in the script file as if it were entered at the prompt.

The rules for script file names are the same as those for MATLAB variable names (see the next Chapter 2, Section 2.1).

As an example, save the compound interest program above in a script file under the name `compint.m`. Then simply enter the name

```
compint
```

at the prompt in the Command Window (as soon as you hit Enter). The statements in `compint.m` will be carried out exactly as if you had pasted them into the Command Window. You have effectively created a new MATLAB command, viz., `compint`.

A script file may be listed in the Command Window with the command `type`, e.g.,

```
type compint
```

(the extension `.m` may be omitted).

Script files provide a useful way of managing large programs which you do not necessarily want to paste into the Command Window every time you run them.

Current directory

When you run a script, you have to make sure that MATLAB's current folder (indicated in the toolbar just above the **Current Folder**) is set to the folder (or directory) in which the script is saved. To change the current folder type the path for the new current folder in the toolbar, or select a folder from the drop-down list of previous working folders, or click on the browse button (it is the first folder with the green arrow that is to the left of the field that indicates the location of the Current Folder). Select a new location for saving and executing files (e.g., if you create files for different courses, you may wish to save your work in folders with the names or numbers of the courses that you are taking).

You can change the current folder from the command line with `cd` command, e.g.,

```
cd \mystuff
```

The command `cd` by itself returns the name of the current directory or current folder (as it is now called in the latest versions of MATLAB).

Running a script from the Current Folder browser

A handy way to run a script is as follows. Select the file in the Current Directory browser. Right-click it. The *context menu* appears (context menus are a general feature of the desktop). Select **Run** from the context menu. The results appear in the Command Window. If you want to edit the script, select **Open** from the context menu.

1.3.3 A program in action

We will now discuss in detail how the compound interest program works.

The MATLAB system is technically called an *interpreter* (as opposed to a *compiler*). This means that each statement presented to the command line is translated (interpreted) into language the computer understands better, and then *immediately* carried out.

A fundamental concept in MATLAB is how numbers are stored in the computer's *random access memory* (RAM). If a MATLAB statement needs to store a number, space in the RAM is set aside for it. You can think of this part of the memory as a bank of boxes or *memory locations*, each of which can hold only one number at a time. These memory locations are referred to by symbolic names in MATLAB statements. So the statement

```
balance = 1000
```

allocates the number 1000 to the memory location named `balance`. Since the contents of `balance` may be changed during a session it is called a *variable*.

The statements in our program are therefore interpreted by MATLAB as follows:

1. Put the number 1000 into variable `balance`.
2. Put the number 0.09 into variable `rate`.
3. Multiply the contents of `rate` by the contents of `balance` and put the answer in `interest`.
4. Add the contents of `balance` to the contents of `interest` and put the answer in `balance`.
5. Display (in the Command Window) the message given in single quotes.
6. Display the contents of `balance`.

It hardly seems necessary to stress this, but these interpreted statements are carried out *in order from the top down*. When the program has finished running, the variables used will have the following values:

```
balance  : 1090
interest : 90
rate     : 0.09
```

Note that the original value of `balance` (1000) is lost.

Try the following exercises:

1. Run the program as it stands.
2. Change the first statement in the program to read

    ```
    balance = 2000;
    ```

 Make sure that you understand what happens now when the program runs.
3. Leave out the line

    ```
    balance = balance + interest;
    ```

 and re-run. Can you explain what happens?
4. Try to rewrite the program so that the original value of `balance` is *not* lost.

A number of questions have probably occurred to you by now, such as

- What names may be used for variables?
- How can numbers be represented?
- What happens if a statement won't fit on one line?
- How can we organize the output more neatly?

These questions will be answered in the next chapter. However, before we write any more complete programs there are some additional basic concepts which need to be introduced. These concepts are introduced in the next chapter.

SUMMARY

- MATLAB is a matrix-based computer system designed to assist in scientific and engineering problem solving.
- To use MATLAB, you enter commands and statements on the command line in the Command Window. They are carried out immediately.
- `quit` or `exit` terminates MATLAB.
- `clc` clears the Command Window.

- `help` and `lookfor` provide help.
- `plot` draws an *x-y* graph in a figure window.
- `grid` draws grid lines on a graph.

EXERCISES

1.1 Give values to variables `a` and `b` on the command line, e.g., `a = 3` and `b = 5`. Write some statements to find the sum, difference, product and quotient of `a` and `b`.

1.2 In Section 1.2.5 of the text a script is given for an animation of the Mexican hat problem. Type this into the editor, save it and execute it. Once you finish debugging it and it executes successfully try modifying it. (a) Change the maximum value of `n` from 3 to 4 and execute the script. (b) Change the time delay in the `pause` function from 0.05 to 0.1. (c) Change the `z=sin(r.*n)./r;` command line to `z=cos(r.*n);` and execute the script.

1.3 Assign a value to the variable `x` on the command line, e.g., `x = 4 * pi^2`. What is the square root of `x`? What is the cosine of the square root of `x`?

1.4 Assign a value to the variable `y` on the command line as follows: `y = -1`. What is the square root of `y`? Show that the answer is

```
ans =
      0 + 1.0000i
```

Yes, MATLAB deals with complex numbers (not just real numbers). Hence the symbol `i` should not be used as an index or as a variable name. By default, it is equal to the square root of -1. (Also, when necessary, `j` is used in MATLAB as a symbol for $\sqrt{-1}$. Hence, it also should not be used as an index or as a variable name.) Give an example of how you have used complex numbers in your studies of mathematics and the sciences up to this point in your education.

Solutions to many of the exercises are in Appendix D.

APPENDIX 1.A SUPPLEMENTARY MATERIAL

Supplementary material related to this chapter can be found online at https://doi.org/10.1016/B978-0-08-102997-8.00007-5.

MATLAB Fundamentals

THE OBJECTIVE OF THIS CHAPTER IS TO INTRODUCE SOME OF THE FUNDAMENTALS OF MATLAB PROGRAMMING, INCLUDING:

- Variables, operators, and expressions
- Arrays (including vectors and matrices)
- Basic input and output
- Repetition (`for`)
- Decisions (`if`)

CONTENTS

Variables 31
Case sensitivity 32

The workspace . 32
Adding commonly used constants to the workspace 33

Arrays: Vectors and matrices.... 34
Initializing vectors: Explicit lists 34
Initializing vectors: The colon operator . 35
The `linspace` *and* `logspace` *functions* . 36
Transposing vectors 37
Subscripts 37
Matrices............... 38
Capturing output ... 39
Structure plan 39

Vertical motion under gravity.... 40

Operators, expressions, and statements 42
Numbers.............. 43
Data types 43
Arithmetic operators 44
Operator precedence........... 44
The colon operator . 45
The transpose operator 46
Arithmetic operations on arrays 46

The tools introduced in this chapter are sufficient to begin solving numerous scientific and engineering problems you may encounter in your course work and in your profession. The last part of this chapter and the next chapter describe an approach to designing reasonably good programs to initiate the building of tools for your own toolbox.

2.1 VARIABLES

Variables are fundamental to programming. In a sense, the art of programming is this:

> Getting the right values in the right variables at the right time

A variable name (like the variable `balance` that we used in Chapter 1) must comply with the following two rules:

- It may consist only of the letters a–z, the digits 0–9, and the underscore (_).
- It must start with a letter.

The name may be as long as you like, but MATLAB only remembers the first 63 characters (to check this on your version, execute the command `namelength-max` in the Command Window of the MATLAB desktop). Examples of valid

Essential MATLAB for Engineers and Scientists. https://doi.org/10.1016/B978-0-08-102997-8.00008-7

Expressions 47
Statements 48
Statements,
commands, and
functions 49
Formula
vectorization 49

Output 52
The disp statement 52
The format
command 54
Scale factors 55

Repeating with
for 55
Square roots with
Newton's method... 56
Factorials! 57
Limit of a sequence 58
The basic for
construct............. 58
for in a single line.. 60
More general for... 60
Avoid for loops by
vectorizing! 60

Decisions 63
The one-line if
statement............ 63
The if-else
construct............. 65
The one-line
if-else statement.. 66
elseif................. 66
Logical operators... 68
Multiple ifs versus
elseif................. 68
Nested ifs 70
Vectorizing ifs?..... 71
The switch
statement............ 71

Complex
numbers 71

Summary 73

Exercises 75

Supplementary
material 81

variable names are r2d2 and pay_day. Examples of invalid names (why?) are pay-day, 2a, name$, and _2a.

A variable is created simply by assigning a value to it at the command line or in a program—for example,

```
a = 98
```

If you attempt to refer to a nonexistent variable you will get the error message in red

```
Undefined function or variable '...'.
```

The official MATLAB documentation refers to all variables as *arrays*, whether they are single-valued (scalars) or multi-valued (vectors or matrices). In other words, a scalar is a 1-by-1 array—an array with a single row and a single column which, of course, is an array of one item.

2.1.1 Case sensitivity

MATLAB is *case-sensitive*, which means it distinguishes between upper- and lowercase letters. Thus, balance, BALANCE and BaLance are three different variables. Many programmers write variable names in lowercase except for the first letter of the second and subsequent words, if the name consists of more than one word run together. This style is known as *camel caps*, the uppercase letters looking like a camel's humps (with a bit of imagination). Examples are camelCaps, milleniumBug, dayOfTheWeek. Some programmers prefer to separate words with underscores.

Command and function names are also case-sensitive. You must *not* use capitals when running built-in functions and commands!

2.2 THE WORKSPACE

Another fundamental concept in MATLAB is the *workspace*. Enter the command clear and then rerun the compound interest program (see Section 1.3.2). Now enter the command who. You should see a list of variables as follows:

```
Your variables are:

balance    interest    rate
```

All the variables you create during a session remain in the workspace until you clear them. You can use or change their values at any stage during the session.

The command who lists the names of all the variables in your workspace. The function ans returns the value of the last expression evaluated but not assigned to a variable. The command whos lists the size of each variable as well:

```
Name            Size          Bytes  Class

balance         1x1               8  double
interest        1x1               8  double
rate            1x1               8  double
```

Each variable here occupies eight *bytes* of storage. A byte is the amount of computer memory required for one character (if you are interested, one byte is the same as eight *bits*). These variables each have a *size* of "1 by 1," because they are *scalars*, as opposed to vectors or matrices (although as mentioned above, MATLAB regards them all as 1-by-1 arrays). The Class double means that the variable holds numeric values as double-precision floating-point (see Section 2.5).

The command clear removes all variables from the workspace. A particular variable can be removed from the workspace (e.g., clear rate). More than one variable can also be cleared (e.g., clear rate balance). Separate the variable names with spaces, *not commas*.

When you run a program, any variables created by it remain in the workspace after it runs. This means that existing variables with the same names are overwritten.

The Workspace browser on the desktop provides a handy visual representation of the workspace. You can view and even change the values of workspace variables with the Array Editor. To activate the Array Editor click on a variable in the Workspace browser or right-click to get the more general context menu. From the context menu you can draw graphs of workspace variables in various ways.

2.2.1 Adding commonly used constants to the workspace

If you often use the same physical or mathematical constants in your MATLAB sessions, you can save them in an M-file and run the file at the start of a session. For example, the following statements can be saved in myconst.m:

```
g = 9.81;               % acceleration due to gravity
avo = 6.023e23;         % Avogadro's number
e = exp(1);             % base of natural log
pi_4 = pi / 4;
log10e = log10( e );
bar_to_kP = 101.325;    % atmospheres to kiloPascals
```

If you run `myconst` at the start of a session, these six variables will be part of the workspace and will be available for the rest of the session or until you `clear` them. This approach to using MATLAB is like a notepad (it is one of many ways). As your experience grows, you will discover many more utilities and capabilities associated with MATLAB's computational and analytical environment.

2.3 ARRAYS: VECTORS AND MATRICES

As mentioned in Chapter 1, the name MATLAB stands for Matrix Laboratory because MATLAB has been designed to work with *matrices*. A matrix is a rectangular object (e.g., a *table*) consisting of rows and columns. We will postpone most of the details of proper matrices and how MATLAB works with them until Chapter 6.

A *vector* is a special type of matrix, having only one row or one column. Vectors are called *lists* or *arrays* in other programming languages. If you haven't come across vectors officially yet, don't worry—just think of them as lists of numbers.

MATLAB handles vectors and matrices in the same way, but since vectors are easier to think about than matrices, we will look at them first. This will enhance your understanding and appreciation of many aspects of MATLAB. As mentioned above, MATLAB refers to scalars, vectors, and matrices generally as arrays. We will also use the term *array* generally, with *vector* and *matrix* referring to the one-dimensional (1D) and two-dimensional (2D) array forms.

2.3.1 Initializing vectors: Explicit lists

As a start, try the accompanying short exercises on the command line. These are all examples of the *explicit list* method of *initializing* vectors. (You won't need reminding about the command prompt ≫ or the need to <**Enter**> any more, so they will no longer appear unless the context absolutely demands it.)

Exercises

2.1 Enter a statement like

```
x = [1 3 0 -1 5]
```

Can you see that you have created a vector (list) with five *elements*? (Make sure to leave out the semicolon so that you can see the list. Also, make sure you hit Enter to execute the command.)

2.2 Enter the command `disp(x)` to see how MATLAB displays a vector.

2.3 Enter the command `whos` (or look in the Workspace browser). Under the heading `Size` you will see that x is 1 by 5, which means 1 row and 5 columns. You will also see that the total number of elements is 5.

2.4 You can use commas instead of spaces between vector elements if you like. Try this:

```
a = [5,6,7]
```

2.5 Don't forget the commas (or spaces) between elements; otherwise, you could end up with something quite different:

```
x = [130 - 15]
```

What do you think this gives? Take the space between the minus sign and 15 to see how the assignment of x changes.

2.6 You can use one vector in a list for another one. Type in the following:

```
a = [1 2 3]
b = [4 5]
c = [a -b]
```

Can you work out what c will look like before displaying it?

2.7 And what about this?

```
a = [1 3 7];
a = [a 0 -1];
```

2.8 Enter the following

```
x = [ ]
```

Note in the Workspace browser that the size of x is given as 0 by 0 because x is *empty*. This means x is defined and can be used where an array is appropriate without causing an error; however, it has no size or value.

Making x empty is *not* the same as saying x = 0 (in the latter case x has size 1 by 1) or clear x (which removes x from the workspace, making it undefined). An empty array may be used to remove elements from an array (see Section 2.3.5).

Remember the following important rules:

- Elements in the list must be enclosed in square brackets, not parentheses.
- Elements in the list must be separated *either by spaces or by commas*.

2.3.2 Initializing vectors: The colon operator

A vector can also be generated (initialized) with the *colon operator*, as we saw in Chapter 1. Enter the following statements:

```
x = 1:10
```

(elements are the integers 1, 2, ... , 10)

```
x = 1:0.5:4
```

(elements are the values 1, 1.5, ... , 4 in increments of 0.5. Note that if the colons separate three values, the *middle* value is the increment);

```
x = 10:-1:1
```

(elements are the integers 10, 9, ... , 1, since the increment is *negative*);

```
x = 1:2:6
```

(elements are 1, 3, 5; note that when the increment is positive but not equal to 1, the last element is not allowed to exceed the value after the second colon);

```
x = 0:-2:-5
```

(elements are 0, −2, −4; note that when the increment is negative but not equal to −1, the last element is not allowed to be less than the value after the second colon);

```
x = 1:0
```

(a complicated way of generating an empty vector!).

2.3.3 The `linspace` and `logspace` functions

The function `linspace` can be used to initialize a vector of equally spaced values:

```
linspace(0, pi/2, 10)
```

creates a vector of 10 equally spaced points from 0 to $\pi/2$ (inclusive).

The function `logspace` can be used to generate logarithmically spaced data. It is a logarithmic equivalent of `linspace`. To illustrate the application of `logspace` try the following:

```
y = logspace(0, 2, 10)
```

This generates the following set of numbers 10 numbers between 10^0 to 10^2 (inclusive): 1.0000, 1.6681, 2.7826, 4.6416, 7.7426, 12.9155, 21.5443, 35.9381, 59.9484, 100.0000. If the last number in this function call is omitted, the number of values of y computed is by default 50. What is the interval between the

numbers 1 and 100 in this example? To compute the distance between the points you can implement the following command:

```
dy = diff(y)
yy = y(1:end-1) + dy./2
plot(yy,dy)
```

You will find that you get a straight line from the point $(yy, dy) = (1.3341, 0.6681)$ to the point $(79.9742, 40.0516)$. Thus, the `logspace` function produces a set of points with an interval between them that increases linearly with y. The variable yy was introduced for two reasons. The first was to generate a vector of the same length as dy. The second was to examine the increase in the interval with increase in y that is obtained with the implementation of `logspace`.

2.3.4 Transposing vectors

All of the vectors examined so far are *row vectors*. Each has one row and several columns. To generate the *column vectors* that are often needed in mathematics, you need to *transpose* such vectors—that is, you need to interchange their rows and columns. This is done with the single quote, or *apostrophe* (`'`), which is the nearest MATLAB can get to the mathematical prime (′) that is often used to indicate the transpose.

Enter `x = 1:5` and then enter `x'` to display the transpose of `x`. Note that `x` itself remains a row vector. Alternatively, or you can create a column vector directly:

```
y = [1 4 8 0 -1]'
```

2.3.5 Subscripts

We can refer to particular elements of a vector by means of *subscripts*. Try the following:

1. Enter `r = rand(1,7)`. This gives you a row vector of seven *random numbers*.
2. Enter `r(3)`. This will display the *third* element of `r`. The numeral 3 is the *subscript*.
3. Enter `r(2:4)`. This should give you the second, third, and fourth elements.
4. What about `r(1:2:7)` and `r([1 7 2 6])`?
5. Use an empty vector to *remove* elements from a vector:

   ```
   r([1 7 2]) = [ ]
   ```

 This will remove elements 1, 7, and 2.

To summarize:

- A subscript is indicated by parentheses.
- A subscript may be a scalar or a vector.
- In MATLAB subscripts always start at 1.
- Fractional subscripts are not allowed.

2.3.6 Matrices

A *matrix* may be thought of as a table consisting of rows and columns. You create a matrix just as you do a vector, except that a semicolon is used to indicate the end of a row. For example, the statement

```
a = [1 2 3; 4 5 6]
```

results in

```
a =
   1     2     3
   4     5     6
```

A matrix may be transposed: With a initialized as above, the statement a' results in

```
ans =
   1     4
   2     5
   3     6
```

A matrix can be constructed from column vectors of the same length. Thus, the statements

```
x = 0:30:180;
table = [x' sin(x*pi/180)']
```

result in

```
table =
         0          0
   30.0000     0.5000
   60.0000     0.8660
   90.0000     1.0000
  120.0000     0.8660
  150.0000     0.5000
  180.0000     0.0000
```

2.3.7 Capturing output

You can use cut and paste techniques to tidy up the output from MATLAB statements if this is necessary for some sort of presentation. Generate the table of angles and sines as shown above. Select all seven rows of numerical output in the Command Window, and copy the selected output to the Editor. You can then edit the output, for example, by inserting text headings above each column (this is easier than trying to get headings to line up over the columns with a `disp` statement). The edited output can in turn be pasted into a report or printed as is (the **File** menu has a number of printing options).

Another way of capturing output is with the `diary` command. The command

```
diary  filename
```

copies everything that subsequently appears in the Command Window to the text file *filename*. You can then edit the resulting file with any text editor (including the MATLAB Editor). Stop recording the session with

```
diary off
```

Note that `diary` *appends* material to an existing file—that is, it adds new information to the end of it.

2.3.8 Structure plan

A structure plan is a top-down design of the steps required to solve a particular problem with a computer. It is typically written in what is called *pseudo-code*—that is, statements in English, mathematics, and MATLAB describing in detail how to solve a problem. You don't have to be an expert in any particular computer language to understand pseudo-code. A structure plan may be written at a number of levels, each of increasing complexity, as the logical structure of the program is developed.

Suppose we want to write a script to convert a temperature on the Fahrenheit scale (where water freezes and boils at 32° and 212°, respectively) to the Celsius scale. A first-level structure plan might be a simple statement of the problem:

1. Initialize Fahrenheit temperature
2. Calculate and display Celsius temperature
3. Stop

Step 1 is pretty straightforward. Step 2 needs elaborating, so the second-level plan could be something like this:

1. Initialize Fahrenheit temperature (*F*)
2. Calculate Celsius temperature (*C*) as follows: Subtract 32 from *F* and multiply by 5/9

3. Display the value of C
4. Stop

There are no hard and fast rules for writing structure plans. The essential point is to cultivate the mental discipline of getting the problem logic clear before attempting to write the program. The top-down approach of structure plans means that the overall structure of a program is clearly thought out before you have to worry about the details of syntax (coding). This reduces the number of errors enormously.

A script to implement this is as follows:

```
% Script file to convert temperatures from F to C
%
F = input(' Temperature in degrees F: ')
       C = (F - 32) * 5 / 9;
disp(['' Temperature in degrees C = ',num2str(C)])
% STOP
```

Two checks of the tool were done. They were for $F = 32$, which gave $C = 0$, and $F = 212$, which gave $C = 100$. The results were found to be correct and hence this simple script is, as such, validated.

The essence of any structure plan and, hence, any computer program can be summarized as follows:

1. Input: Declare and assign input variables.
2. Operations: Solve expressions that use the input variables.
3. Output: Display in graphs or tables the desired results.

2.4 VERTICAL MOTION UNDER GRAVITY

If a stone is thrown vertically upward with an initial speed u, its vertical displacement s after an elapsed time t is given by the formula $s = ut - gt^2/2$, where g is the acceleration due to gravity. Air resistance is ignored. We would like to compute the value of s over a period of about 12.3 seconds at intervals of 0.1 seconds, and plot the distance–time graph over this period, as shown in Fig. 2.1. The structure plan for this problem is as follows:

```
1.    % Assign the data (g, u, and t) to MATLAB variables
2.    % Calculate the value of s according to the formula
3.    % Plot the graph of s against t
4.    % Stop
```

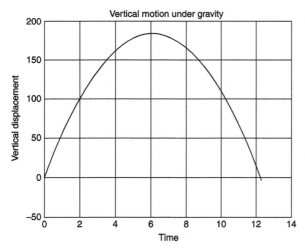

Vertical motion under gravity

FIGURE 2.1
Distance-time graph of a stone thrown vertically upward.

This plan may seem trivial and a waste of time to write down. Yet you would be surprised how many beginners, preferring to rush straight to the computer, start with step 2 instead of step 1. It is well worth developing the mental discipline of structure-planning your program first. You can even use cut and paste to plan as follows:

1. Type the structure plan into the Editor (each line preceded by % as shown above).
2. Paste a second copy of the plan directly below the first.
3. Translate each line in the second copy into a MATLAB statement or statements (add % comments as in the example below).
4. Finally, paste all the translated MATLAB statements into the Command Window and run them (or you can just click on the green triangle in the toolbar of the Editor to execute your script).
5. If necessary, go back to the Editor to make corrections and repaste the corrected statements to the Command Window (or save the program in the Editor as an M-file and execute it).

You might like to try this as an exercise before looking at the final program, which is as follows:

```
% Vertical motion under gravity
g = 9.81; % acceleration due
% to gravity
u = 60; % initial velocity in
% meters/sec
t = 0 : 0.01 : 12.3; % time in seconds
```

```
s = u * t - g / 2 * t .^ 2; % vertical displacement
% in meters
plot(t, s,'k','LineWidth',3)
title( 'Vertical motion under gravity' )
xlabel( 'time' ), ylabel( 'vertical displacement' )
grid
```

The graphical output is shown in Fig. 2.1.

Note the following points:

- Anything in a line following the symbol % is ignored by MATLAB and may be used as a comment (description).
- The statement t = 0 : 0.1 : 12.3 sets up a vector.
- The formula for s is evaluated *for every element* of the vector t, making another vector.
- The expression t.^2 squares *each element* in t. This is called an *array operation* and is different from squaring the vector itself, which is a *matrix* operation, as we will see later.
- More than one statement can be entered on the same line if the statements are separated by commas.
- A statement or group of statements can be continued to the next line with an *ellipsis* of three or more dots (. . .).
- The statement disp([t' s']) first transposes the row vectors t and s into two columns and constructs a matrix from them, which is then displayed.

You might want to save the program under a helpful name, like throw.m, if you think you might come back to it. In that case, it would be worth keeping the structure plan as part of the file; just insert % symbols in front of each line. This way, the plan reminds you what the program does when you look at it again after some months. Note that you can use the context menu in the Editor to **Comment/Uncomment** a block of selected text. After you block selected text, right-click to see the context menu. To comment the text, scroll down to **Comment**, point, and click.

2.5 OPERATORS, EXPRESSIONS, AND STATEMENTS

Any program worth its salt actually does something. What it basically does is evaluate *expressions*, such as

```
u*t - g/2*t.^2
```

and execute (carry out) statements, such as

```
balance = balance + interest
```

MATLAB is described as an *expression* based language because it interprets and evaluates typed expressions. Expressions are constructed from a variety of things, such as numbers, variables, and operators. First we need to look at numbers.

2.5.1 Numbers

Numbers can be represented in MATLAB in the usual decimal form (*fixed point*) with an optional decimal point,

```
1.2345    -123    .0001
```

They may also be represented in *scientific notation*. For example, 1.2345×10^9 may be represented in MATLAB as `1.2345e9`. This is also called *floating-point* notation. The number has two parts: the *mantissa*, which may have an optional decimal point (`1.2345` in this example) and the *exponent* (`9`), which must be an integer (signed or unsigned). Mantissa and exponent must be separated by the letter `e` (or `E`). The mantissa is multiplied by the power of 10 indicated by the exponent.

Note that the following is *not* scientific notation: `1.2345*10^9`. It is actually an *expression* involving two arithmetic operations (`*` and `^`) and therefore more time consuming. Use scientific notation if the numbers are very small or very large, since there's less chance of making a mistake (e.g., represent 0.000000001 as `1e-9`).

On computers using standard floating-point arithmetic, numbers are represented to approximately 16 significant decimal digits. The *relative accuracy* of numbers is given by the function `eps`, which is defined as the distance between 1.0 and the next largest floating-point number. Enter `eps` to see its value on your computer.

The range of numbers is roughly $\pm 10^{-308}$ to $\pm 10^{308}$. Precise values for your computer are returned by the MATLAB functions `realmin` and `realmax`.

As an exercise, enter the following numbers at the command prompt in scientific notation (answers follow in parentheses):

1.234×10^5, -8.765×10^{-4}, 10^{-15}, -10^{12}
($1.234e+05$, $-8.765e-04$, $1e-15$, $-1e+12$)

2.5.2 Data types

MATLAB has more than a dozen fundamental data types (or classes). The default numeric data type is double precision; all MATLAB computations are in double precision. More information on data types can be found in the Help index.

Table 2.1 Arithmetic operations between two scalars.

Operation	Algebraic form	MATLAB
Addition	$a + b$	a + b
Subtraction	$a - b$	a - b
Multiplication	$a \times b$	a * b
Right division	a/b	a / b
Left division	b/a	a \ b
Power	a^b	a ^ b

Table 2.2 Precedence of arithmetic operations.

Precedence	Operator
1	Parentheses (round brackets)
2	Power, left to right
3	Multiplication and division, left to right
4	Addition and subtraction, left to right

MATLAB also supports signed and unsigned integer types and single-precision floating-point, by means of functions such as `int8`, `uint8`, `single`, and the like. However, before mathematical operations can be performed on such types, they must be converted to double precision using the `double` function.

2.5.3 Arithmetic operators

The evaluation of expressions is achieved by means of *arithmetic operators*. The arithmetic operations on two *scalar* constants or variables are shown in Table 2.1. Operators operate on *operands* (a and b in the table).

Left division seems a little curious: Divide the right operand by the *left* operand. For *scalar* operands the expressions `1/3` and `3\1` have the same numerical value (a colleague of mine speaks of the latter as "3 under 1"). However, *matrix* left division has an entirely different meaning, as we will see later.

2.5.4 Operator precedence

Several operations may be combined in one expression—for example, `g * t ^ 2`. MATLAB has strict precedence rules for which operations are performed first in such cases. The precedence rules for the operators in Table 2.1 are shown in Table 2.2. Note that parentheses have the highest precedence. Note also the difference between parentheses and square brackets. The former are used to alter the precedence of operators and to denote subscripts, while the latter are used to create vectors.

When operators in an expression have the same precedence, the operations are carried out from left to right. So a / b * c is evaluated as (a / b) * c and *not* as a / (b * c).

Exercises

2.1 Evaluate the following MATLAB expressions yourself before checking the answers in MATLAB:

```
1 + 2 * 3
4 / 2 * 2
1 + 2 / 4
1 + 2 \ 4
2 * 2 ^ 3
2 * 3 \ 3
2 ^ (1 + 2)/3
1/2e-1
```

2.2 Use MATLAB to evaluate the following expressions. Answers are in parentheses.

(a) $\frac{1}{2 \times 3}$ (0.1667)

(b) $2^{2 \times 3}$ (64)

(c) $1.5 \times 10^{-4} + 2.5 \times 10^{-2}$ (0.0252; use scientific or floating-point notation)

2.5.5 The colon operator

The colon operator has a lower precedence than the plus operator, as the following shows:

```
1+1:5
```

The addition is carried out first and a vector with elements 2, ... , 5 is then initialized.

You may be surprised at the following:

```
1+[1:5]
```

Were you? The value 1 is added to *each element* of the vector 1:5. In this context, the addition is called an *array operation* because it operates on each element of the vector (array). Array operations are discussed below.

See Appendix A for a complete list of MATLAB operators and their precedences.

Table 2.3 Arithmetic operators that operate element-by-element on arrays.

Operator	Description
.*	Multiplication
./	Right division
.\	Left division
.^	Power

2.5.6 The transpose operator

The transpose operator has the highest precedence. Try

```
1:5'
```

The 5 is transposed first (into itself since it is a scalar!), and then a row vector is formed. Use square brackets if you want to transpose the whole vector:

```
[1:5]'
```

2.5.7 Arithmetic operations on arrays

Enter the following statements at the command line:

```
a = [2 4 5];
b = [6 2 2];
a .* b
a ./ b
```

MATLAB has four additional arithmetic operators, as shown in Table 2.3 that work on corresponding elements of arrays with equal dimensions. They are sometimes called *array* or *element-by-element* operations because they are performed element by element. For example, a .* b results in the following vector (sometimes called the *array product*):

```
[a(1)*b(1) a(2)*b(2) a(3)*b(3)]
```

that is, [12 8 10].

You will have seen that a ./ b gives element-by-element division. Now try [2 3 4] .^ [4 3 1]. The ith element of the first vector is raised to the power of the ith element of the second vector. The period (dot) is necessary for the array operations of multiplication, division, and exponentiation because these operations are defined differently for matrices; they are then called *matrix operations* (see Chapter 6). With a and b as defined above, try a + b and a - b.

For addition and subtraction, array operations and matrix operations are the same, so we don't need the period to distinguish them.

When array operations are applied to two vectors, both vectors must be the same size!

Array operations also apply between a scalar and a nonscalar. Check this with 3 .* a and a .^ 2. This property is called *scalar expansion*. Multiplication and division operations between scalars and nonscalars can be written with or without the period (i.e., if a is a vector, 3 .* a is the same as 3 * a).

A common application of element-by-element multiplication is finding the *scalar product* (also called the *dot* product) of two vectors **x** and **y**, which is defined as

$$\mathbf{x} \cdot \mathbf{y} = \sum_i x_i y_i$$

The MATLAB function sum(z) finds the sum of the elements of the vector **z**, so the statement sum(a .* b) will find the scalar product of a and b (30 for a and b defined above).

Exercises

Use MATLAB array operations to do the following:

2.1 Add 1 to each element of the vector [2 3 -1].
2.2 Multiply each element of the vector [1 4 8] by 3.
2.3 Find the array product of the two vectors [1 2 3] and [0 -1 1]. (Answer: [0 -2 3])
2.4 Square each element of the vector [2 3 1].

2.5.8 Expressions

An *expression* is a formula consisting of variables, numbers, operators, and function names. It is evaluated when you enter it at the MATLAB prompt. For example, evaluate 2π as follows:

```
2 * pi
```

MATLAB's response is

```
ans =
    6.2832
```

Note that MATLAB uses the function ans (which stands for answer) to return the last expression to be evaluated but not assigned to a variable.

If an expression is terminated with a semicolon (;), its value is not displayed, although it is still returned by `ans`.

2.5.9 Statements

MATLAB *statements* are frequently of the form

variable = *expression*

as in

```
s = u * t - g / 2 * t .^ 2;
```

This is an example of an *assignment* statement because the value of the expression on the *right* is *assigned* to the variable (`s`) on the *left*. Assignment always works in this direction. Note that the object on the left-hand side of the assignment must be a variable name. A common mistake is to get the statement the wrong way around, as in

```
a + b = c
```

Basically any line that you enter in the Command Window or in a program, which MATLAB accepts, is a statement, so a statement can be an assignment, a command, or simply an expression, such as

```
x = 29;          % assignment
clear            % command
pi/2             % expression
```

This naming convention is in keeping with most programming languages and serves to emphasize the different types of statements that are found in programming. However, the MATLAB documentation tends to refer to all of these as "functions."

As we have seen, a semicolon at the end of an assignment or expression suppresses any output. This is useful for suppressing irritating output of intermediate results (or large matrices).

A statement that is too long to fit on one line may be continued to the next line with an *ellipsis* of at least three dots:

```
x = 3 * 4 - 8 ....
    / 2 ^ 2;
```

Statements on the same line may be separated by commas (output not suppressed) or semicolons (output suppressed):

```
a = 2; b = 3, c = 4;
```

Note that the commas and semicolons are not technically part of the statements; they are *separators*.

Statements may involve array operations, in which case the variable on the left-hand side may become a vector or a matrix.

2.5.10 Statements, commands, and functions

The distinction between MATLAB *statements*, *commands*, and *functions* can be a little fuzzy, since all can be entered on the command line. However, it is helpful to think of commands as changing the general environment in some way, for example, `load`, `save`, and `clear`. Statements do the sort of thing we usually associate with programming, such as evaluating expressions and carrying out assignments, making decisions (`if`), and repeating (`for`). Functions return with calculated values or perform some operation on data, such as `sin` and `plot`.

2.5.11 Formula vectorization

With array operations, you can easily evaluate a formula repeatedly for a large set of data. This is one of MATLAB's most useful and powerful features, and you should always look for ways to exploit it.

Let us again consider, as an example, the calculation of compound interest. An amount of money A invested over a period of years n with an annual interest rate r grows to an amount $A(1 + r)^n$. Suppose we want to calculate final balances for investments of $750, $1000, $3000, $5000, and $11,999 over 10 years with an interest rate of 9%. The following program (`comp.m`) uses array operations on a vector of initial investments to do this:

```
format bank
A = [750 1000 3000 5000 11999];
r = 0.09;
n = 10;
B = A * (1 + r) ^ n;
disp( [A' B'] )
```

The output is

```
   750.00      1775.52
  1000.00      2367.36
  3000.00      7102.09
  5000.00     11836.82
 11999.00     28406.00
```

Note the following:

- In the statement B = A * (1 + r) ^ n, the expression (1 + r) ^ n is evaluated first because exponentiation has a higher precedence than multiplication.
- Each element of the vector A is then multiplied by the scalar (1 + r) ^ n (scalar expansion).
- The operator * may be used instead of .* because the multiplication is between a scalar and a nonscalar (although .* would not cause an error because a scalar is a special case of an array).
- A table is displayed whose columns are given by the transposes of A and B.

This process is called formula *vectorization*. The operation in the statement described in bullet item 1 is such that every element in the vector B is determined by operating on every element of vector A all at once, by interpreting once a single command line.

See if you can adjust the program comp.m to find the balances for a single amount *A* ($1000) over 1, 5, 10, 15, and 20 years. (*Hint*: use a vector for n: [1 5 10 15 20].)

Exercises

2.1 Evaluate the following expressions yourself (before you use MATLAB to check). The numerical answers are in parentheses.
 - (a) 2 / 2 * 3 (3)
 - (b) 2 / 3 ^ 2 (2/9)
 - (c) (2 / 3) ^ 2 (4/9)
 - (d) 2 + 3 * 4 - 4 (10)
 - (e) 2 ^ 2 * 3 / 4 + 3 (6)
 - (f) 2 ^ (2 * 3) / (4 + 3) (64/7)
 - (g) 2 * 3 + 4 (10)
 - (h) 2 ^ 3 ^ 2 (64)
 - (i) - 4 ^ 2 (-16; ^ has higher precedence than -)

2.2 Use MATLAB to evaluate the following expressions. The answers are in parentheses.
 - (a) $\sqrt{2}$ (1.4142; use sqrt or ^0.5)
 - (b) $\dfrac{3+4}{5+6}$ (0.6364; use brackets)
 - (c) Find the sum of 5 and 3 divided by their product (0.5333)
 - (d) 2^{3^2} (512)
 - (e) Find the square of 2π (39.4784; use pi)
 - (f) $2\pi^2$ (19.7392)
 - (g) $1/\sqrt{2\pi}$ (0.3989)
 - (h) $\dfrac{1}{2\sqrt{\pi}}$ (0.2821)

(i) Find the cube root of the product of 2.3 and 4.5 (2.1793)

(j) $\dfrac{1 - \frac{2}{3+2}}{1 + \frac{2}{3-2}}$ (0.2)

(k) $1000(1 + 0.15/12)^{60}$ (2107.2—for example, $1000 deposited for 5 years at 15% per year, with the interest compounded monthly)

(l) $(0.0000123 + 5.678 \times 10^{-3}) \times 0.4567 \times 10^{-4}$ (2.5988×10^{-7}; use scientific notation—for example, `1.23e-5` ... ; do *not* use ^)

2.3 Try to avoid using unnecessary parentheses in an expression. Can you spot the errors in the following expression? (Test your corrected version with MATLAB.)

```
(2(3+4)/(5*(6+1))^2
```

Note that the MATLAB Editor has two useful ways of dealing with the problem of "unbalanced delimiters" (which you should know about if you have been working through Help!):

- When you type a closing delimiter, that is, a right),], or), its matching opening delimiter is briefly highlighted. So if you don't see the highlighting when you type a right delimiter, you immediately know you've got one too many.

2.4 Set up a vector n with elements 1, 2, 3, 4, 5. Use MATLAB array operations on it to set up the following four vectors, each with five elements:

(a) 2, 4, 6, 8, 10
(b) 1/2, 1, 3/2, 2, 5/2
(c) 1, 1/2, 1/3, 1/4, 1/5
(d) $1, 1/2^2, 1/3^2, 1/4^2, 1/5^2$

2.5 Suppose vectors a and b are defined as follows:

```
a = [2 -1 5 0];
b = [3 2 -1 4];
```

Evaluate by hand the vector c in the following statements. Check your answers with MATLAB.

(a) `c = a - b;`
(b) `c = b + a - 3;`
(c) `c = 2 * a + a .^ b;`
(d) `c = b ./ a;`
(e) `c = b . a;`
(f) `c = a .^ b;`
(g) `c = 2.^b+a;`
(h) `c = 2*b/3.*a;`
(i) `c = b*2.*a;`

2.6 Water freezes at 32° and boils at 212° on the Fahrenheit scale. If *C* and *F* are Celsius and Fahrenheit temperatures, the formula

$$F = 9C/5 + 32$$

converts from one to the other. Use the MATLAB command line to convert Celsius 37° (normal human temperature) to Fahrenheit (98.6°).

2.7 Engineers often have to convert from one unit of measurement to another, which can be tricky sometimes. You need to think through the process carefully. For example, convert 5 acres to hectares, given that an acre is 4840 square yards, a yard is 36 inches, an inch is 2.54 centimeters, and a hectare is 10,000 square meters. The best approach is to develop a formula to convert x acres to hectares. You can do this as follows.

- One square yard $= (36 \times 2.54)^2$ cm^2
- So one acre $= 4840 \times (36 \times 2.54)^2$ cm^2
 $$= 0.4047 \times 10^8 \text{ cm}^2$$
 $$= 0.4047 \text{ hectares}$$
- So x acres $= 0.4047 \times x$ hectares

Once you have found the formula (but not before), MATLAB can do the rest:

```
x = 5;            % acres
h = 0.4047 * x;   % hectares
disp( h )
```

2.8 Develop formulae for the following conversions, and use some MATLAB statements to find the answers (in parentheses).
 (a) Convert 22 yards (an imperial cricket pitch) to meters. (20.117 meters)
 (b) One pound (weight) = 454 grams. Convert 75 kilograms to pounds. (165.20 pounds)
 (c) Convert 49 meters/second (terminal velocity for a falling human-shaped object) to kilometers per hour. (176.4 kilometers per hour)
 (d) One atmosphere pressure = 14.7 pounds per square inch (psi) = 101.325 kilo Pascals (kPa). Convert 40 psi to kPa. (275.71 kPa)
 (e) One calorie = 4.184 joules. Convert 6.25 kilojoules to calories. (1.494 kilocalories)

2.6 OUTPUT

There are two straightforward ways of getting output from MATLAB:

- Entering a variable name, assignment, or expression on the command line, without a semicolon
- Using the `disp` statement (e.g., `disp( x )`)

2.6.1 The `disp` statement

The general form of `disp` for a numeric variable is

```
disp( variable )
```

When you use disp, the variable name is not displayed, and you don't get a line feed before the value is displayed, as you do when you enter a variable name on the command line without a semicolon. disp generally gives a neater display.

You can also use disp to display a message enclosed in apostrophes (called a *string*). Apostrophes that are part of the message must be repeated:

```
disp( 'Pilate said, ''What is truth?''' );
```

To display a message and a value on the same line, use the following trick:

```
x = 2;
disp( ['The answer is ', num2str(x)] );
```

The output should be

```
The answer is 2
```

Square brackets create a vector, as we have seen. If we want to display a *string*, we create it; that is, we type a message between apostrophes. This we have done already in the above example by defining the string 'The answer is '. Note that the last space before the second apostrophe is part of the string. *All* the components of a MATLAB array must be either numbers or strings (unless you use a *cell array*—see Chapter 10), so we convert the number x to its *string representation* with the function num2str; read this as "number to string."

You can display more than one number on a line as follows:

```
disp( [x y z] )
```

The square brackets create a vector with three elements, which are all displayed.

The command more on enables *paging* of output. This is very useful when displaying large matrices, for example, rand(100000,7) (see help more for details). If you forget to switch on more before displaying a huge matrix, you can always stop the display with **Ctrl+C**.

As you gain experience with MATLAB you may wish to learn more about the input and output capabilities in MATLAB. You can begin your search for information by clicking the question mark at the top of the desktop to open the help documents. Then search for fopen, a utility which allows you to open a file. Scroll to the bottom of the page in the help manual on this topic and find the following list of functions: fclose, feof, ferror, fprintf, fread, fscanf, fseek, ftell, fwrite. Click on fprintf, which is a formatted output

utility that is popular if you are a C-language programmer. Search for `input` in the help manual to learn more about this function that is used in a number of examples in this text. Of course, the simplest input of data is the assignment of values to variables in a program of commands.

2.6.2 The `format` **command**

The term *format* refers to how something is laid out: in this case MATLAB output. The default format in MATLAB has the following basic output rules:

- It always attempts to display integers (whole numbers) exactly. However, if the integer is too large, it is displayed in scientific notation with five significant digits—1234567890 is displayed as `1.2346e+009` (i.e., 1.2346×10^9). Check this by first entering 123456789 at the command line and then 1234567890.
- Numbers with decimal parts are displayed with four significant digits. If the value x is in the range $0.001 < x \leq 1000$, it is displayed in fixed-point form; otherwise, scientific (floating-point) notation is used, in which case the *mantissa* is between 1 and 9.9999 (e.g., 1000.1 is displayed as `1.0001e+003`). Check this by entering the following numbers at the prompt (on separate lines): 0.0011, 0.0009, 1/3, 5/3, 2999/3, 3001/3.

You can change from the default with variations on the `format` command, as follows. If you want values displayed in scientific notation (floating-point form) whatever their size, enter the command

```
format short e
```

All output from subsequent `disp` statements will be in scientific notation, with five significant digits, until the next `format` command is issued. Enter this command and check it with the following values: 0.0123456, 1.23456, 123.456 (all on separate lines).

If you want more accurate output, you can use `format long e`. This also gives scientific notation but with 15 significant digits. Try it out on 1/7. Use `format long` to get fixed-point notation with 15 significant digits. Try 100/7 and `pi`. If you're not sure of the order of magnitude of your output you can try `format short g` or `format long g`. The g stands for "general." MATLAB decides in each case whether to use fixed or floating point.

Use `format bank` for financial calculations; you get fixed point with two decimal digits (for cents). Try it on 10000/7. Suppress irritating line feeds with `format compact`, which gives a more compact display. `format loose` reverts to a more airy display. Use `format hex` to get hexadecimal display.

Use `format rat` to display a number as a rational approximation (ratio of two integers). For example, `pi` is displayed as 355/113, a pleasant change from the

tired old 22/7. Note that even this is an approximation! Try out `format rat` on $\sqrt{2}$ and e (`exp(1)`).

The symbols +, −, and a space are displayed for positive, negative, and zero elements of a vector or matrix after the command `format +`. In certain applications this is a convenient way of displaying matrices. The command `format` by itself reverts to the default format. You can always try `help format` if you're confused!

2.6.3 Scale factors

Enter the following commands (MATLAB's response is also shown).

```
>> format compact
>> x = [1e3 1 1e-4]
x =
   1.0e+003 *
     1.0000    0.0010    0.0000
```

With `format short` (the default) and `format long`, a common *scale factor* is applied to the whole vector if its elements are very large or very small or differ greatly in magnitude. In this example, the common scale factor is 1000, so the elements displayed must all be multiplied by it to get their proper value—for example, for the second element 1.0e+003 * 0.0010 gives 1. Taking a factor of 1000 out of the third element (`1e-4`) leaves `1e-7`, which is represented by 0.0000 since only four decimal digits are displayed.

If you don't want a scale factor, try `format bank` or `format short e`:

```
>> x
x =
   1000.00          1.00          0.00
>> format short e
>> x
x =
   1.0000e+003   1.0000e+000   1.0000e-004
```

2.7 REPEATING WITH for

So far we have seen how to get data into a program (i.e., provide *input*), how to do arithmetic, and how to get some results (i.e., get *output*). In this section we look at a new feature: repetition. This is implemented by the extremely

powerful `for` construct. We will first look at some examples of its use, followed by explanations.

For starters, enter the following group of statements on the command line. Enter the command `format compact` first to make the output neater:

```
for i = 1:5, disp(i), end
```

Now change it slightly to

```
for i = 1:3, disp(i), end
```

And what about

```
for i = 1:0, disp(i), end
```

Can you see what's happening? The `disp` statement is repeated five times, three times, and not at all.

2.7.1 Square roots with Newton's method

The square root x of any positive number a may be found using only the arithmetic operations of addition, subtraction, and division with *Newton's method*. This is an iterative (repetitive) procedure that refines an initial guess.

To introduce in a rather elementary way the notion of *structured programming* (to be described in more detail in Chapter 3), let us consider the structure plan of the algorithm to find a square root and a program with sample output for $a = 2$.

Here is the structure plan:

1. Initialize a
2. Initialize x to $a/2$
3. Repeat 6 times (say) the following:
 Replace x by $(x + a/x)/2$
 Display x
4. Stop

Here is the program:

```
a = 2;
x = a/2;
disp(['The approach to sqrt(a) for a = ', num2str(a)])
```

```
for i = 1:6
   x = (x + a / x) / 2;
   disp( x )
end

disp( 'Matlab''s value: ' )
disp( sqrt(2) )
```

Here is the output (after selecting `format long`):

```
The approach to sqrt(a) for a = 2
   1.50000000000000
   1.41666666666667
   1.41421568627451
   1.41421356237469
   1.41421356237310
   1.41421356237310

Matlab's value:
   1.41421356237310
```

The value of x converges to a limit rather quickly in this case, $\sqrt{a}$. Note that it is identical to the value returned by MATLAB's `sqrt` function. Most computers and calculators use a similar method internally to compute square roots and other standard mathematical functions.

The general form of Newton's method is presented in Chapter 17.

2.7.2 Factorials!

Run the following program to generate a list of n and $n!$ ("n factorial," or "n shriek"), where

$$n! = 1 \times 2 \times 3 \times \ldots \times (n-1) \times n \qquad (2.1)$$

```
n = 10;
fact = 1;
for k = 1:n
    fact = k * fact;
    disp( [k fact] )
end
```

Experiment to find the largest value of n for which MATLAB can find the n factorial. (You had better leave out the `disp` statement! Or you can move it from above the `end` command to below it.)

2.7.3 Limit of a sequence

for loops are ideal for computing successive members of a sequence (as in Newton's method). The following example also highlights a problem that sometimes occurs when computing a limit. Consider the sequence

$$x_n = \frac{a^n}{n!}, \quad n = 1, 2, 3 \ldots$$

where a is any constant and $n!$ is the factorial function defined above. The question is this: What is the limit of this sequence as n gets indefinitely large? Let's take the case $a = 10$. If we try to compute x_n directly, we can get into trouble, because $n!$ grows very rapidly as n increases, and numerical *overflow* can occur. However, the situation is neatly transformed if we spot that x_n is related to x_{n-1} as follows:

$$x_n = \frac{a x_{n-1}}{n}.$$

There are no numerical problems now.

The following program computes x_n for $a = 10$ and increasing values of n.

```
a = 10;
x = 1;
k = 20;              % number of terms

for n = 1:k
  x = a * x / n;
  disp( [n x] )
end
```

2.7.4 The basic for construct

In general the most common form of the for loop (for use in a program, not on the command line) is

```
for index = j:k
    statements;
end
```

or

```
for index = j:m:k
    statements;
end
```

Note the following points carefully:

- $j:k$ is a vector with elements $j, j+1, j+2, \ldots, k$.
- $j:m:k$ is a vector with elements $j, j+m, j+2m, \ldots$, such that the last element does not exceed k if $m > 0$ or is not less than k if $m < 0$.
- *index* must be a variable. Each time through the loop it will contain the next element of the vector $j:k$ or $j:m:k$, and *statements* (there may be one or more) are carried out for each of these values.
 If the for construct has the form

```
for k = first:increment:last
```

The number of times the loop is executed may be calculated from the following equation:

$$\texttt{floor} \left(\frac{\texttt{last} - \texttt{first}}{\texttt{increment}} \right) + 1$$

where the MATLAB function `floor(x)` rounds x *down* toward $-\infty$. This value is called the *iteration* or *trip count*. As an example, let us consider the statement for i = 1:2:6. It has an iteration count of

$$\texttt{floor} \left(\frac{6-1}{2} \right) + 1 = \texttt{floor} \left(\frac{5}{2} \right) + 1 = 3$$

Thus i takes the values 1, 3, 5. Note that if the iteration count is negative, the loop is not executed.
- On completion of the for loop the index contains the last value used.
- If the vector $j:k$ or $j:m:k$ is empty, *statements* are not executed and control passes to the statement following end.
- The index does not have to appear explicitly in *statements*. It is basically a counter. In fact, if the index *does* appear explicitly in *statements*, the for can often be *vectorized* (more details on this are given in Section 2.7.7). A simple example of a more efficient (faster) program is as follows. The examples with disp at the beginning of this section were for illustration only; strictly, it would be more efficient to say (without "for")

```
i = 1:5; disp( i' )
```

Can you see the difference? In this case i is assigned as a vector (hence, this change *vectorizes* the original program).
- It is good programming style to *indent* (tabulate) the statements inside a for loop. You may have noticed that the Editor does this for you automatically with a feature called *smart indenting*.

2.7.5 `for` **in a single line**

If you insist on using `for` in a single line, here is the general form:

```
for index = j:k; statements; end
```

or

```
for index = j:m:k; statements; end
```

Note the following:

- Don't forget the commas (semicolons will also do if appropriate). If you leave them out you will get an error message.
- Again, *statements* can be one or more statements separated by commas or semicolons.
- If you leave out `end`, MATLAB will wait for you to enter it. Nothing will happen until you do so.

2.7.6 **More general** `for`

A more general form of `for` is

```
for index = v
```

where `v` is any vector. The index moves through each element of the vector in turn, providing a neat way of processing each item in a list. Other forms of the `for` loop as well as the `while` loop will be discussed in Chapter 8.

2.7.7 **Avoid** `for` **loops by vectorizing!**

There are situations where a `for` loop is essential, as in many of the examples in this section so far. However, given the way MATLAB has been designed, `for` loops tend to be inefficient in terms of computing time. If you have written a `for` loop that involves the index of the loop in an expression, it may be possible to vectorize the expression, making use of array operations where necessary, as the following examples show.

Suppose you want to evaluate

$$\sum_{n=1}^{1\,00\,000} n$$

(and can't remember the formula for the sum). Here's how to do it with a `for` loop (run the program, which also times how long it takes):

```
t0 = clock;
s = 0;
for n = 1:100000
    s = s + n;
end
etime(clock, t0)
```

The MATLAB function `clock` returns a six-element vector with the current date and time in the format year, month, day, hour, minute, seconds. Thus, `t0` records when the calculation starts.

The function `etime` returns the time in seconds elapsed between its two arguments, which must be vectors as returned by `clock`. On a Pentium II, it returned about 3.35 seconds, which is the total time for this calculation. (If you have a faster PC, it should take less time.)

Now try to vectorize this calculation (before looking at the solution). Here it is:

```
t0 = clock;
n = 1:100000;
s = sum( n );
etime(clock, t0)
```

This way takes only 0.06 seconds on the same PC—more than 50 times faster!

There is a neater way of monitoring the time taken to interpret MATLAB statements: the `tic` and `toc` function. Suppose you want to evaluate

$$\sum_{n=1}^{100\,000} \frac{1}{n^2}$$

Here's the `for` loop version:

```
tic
s = 0;
for n = 1:100000
    s = s + 1/n^2;
end
toc
```

which takes about 6 seconds on the same PC. Once again, try to vectorize the sum:

```
tic
n = 1:100000;
s = sum( 1./n.^2 );
toc
```

The same PC gives a time of about 0.05 seconds for the vectorized version—more than 100 times faster! (Of course, the computation time in these examples is small regardless of the method applied. However, learning how to improve the efficiency of computation to solve more complex scientific or engineering problems will be helpful as you develop good programming skills. More details on good problem-solving and program design practices are introduced at the end of this chapter and dealt with, in more detail, in the next.)

Series with alternating signs are a little more challenging. This series sums to ln(2) (the natural logarithm of 2):

$$1 - \frac{1}{2} + \frac{1}{3} - \frac{1}{4} + \frac{1}{5} - \cdots$$

Here's how to find the sum of the first 9999 terms with a `for` loop (note how to handle the alternating sign):

```
sign = -1;
s = 0;

for n = 1:9999
sign = -sign;
    s = s + sign / n;
end
```

Try it. You should get 0.6932. MATLAB's `log(2)` gives 0.6931. Not bad.

The vectorized version is as follows:

```
n = 1:2:9999;
s = sum( 1./n - 1./(n+1) )
```

If you time the two versions, you will again find that the vectorized form is many times faster.

MATLAB's functions naturally exploit vectorization wherever possible. For example, `prod(1:n)` will find $n!$ much faster than the code at the beginning of this section (for large values of n).

Exercises

Write MATLAB programs to find the following sums with `for` loops and by vectorization. Time both versions in each case.

- $1^2 + 2^2 + 3^2 + \cdots + 1\,000^2$ (sum is 333 833 500)
- $1 - \dfrac{1}{3} + \dfrac{1}{5} - \dfrac{1}{7} + \dfrac{1}{9} - \cdots - \dfrac{1}{1003}$ (sum is 0.7849—converges slowly to $\pi/4$)
- Sum the left-hand side of the series

$$\frac{1}{1^2 \cdot 3^2} + \frac{1}{3^2 \cdot 5^2} + \frac{1}{5^2 \cdot 7^2} + \cdots = \frac{\pi^2 - 8}{16}$$ (sum is 0.1169—with 500 terms)

2.8 DECISIONS

The MATLAB function `rand` generates a random number in the range 0–1. Enter the following two statements at the command line:

```
r = rand
if r > 0.5 disp( 'greater indeed' ), end
```

MATLAB should only display the message `greater indeed` if r is in fact greater than 0.5 (check by displaying r). Repeat a few times—cut and paste from the Command History window (make sure that a new r is generated each time).

As a slightly different but related exercise, enter the following *logical expression* on the command line:

```
2 > 0
```

Now enter the logical expression `-1 > 0`. MATLAB gives a value of 1 to a logical expression that is *true* and 0 to one that is *false*.

2.8.1 The one-line `if` statement

In the last example MATLAB has to make a decision; it must decide whether or not r is greater than 0.5. The `if` construct, which is fundamental to all computing languages, is the basis of such decision making. The simplest form of `if` in a single line is

```
if condition; statements; end
```

Note the following points:

Table 2.4 Relational operators.

Relational operator	Meaning
<	less than
<=	less than or equal
==	equal
~=	not equal
>	greater than
>=	greater than or equal

- *condition* is usually a *logical expression* (i.e., it contains a *relational operator*), which is either *true* or *false*. The relational operators are shown in Table 2.4. MATLAB allows you to use an arithmetic expression for *condition*. If the expression evaluates to 0, it is regarded as false; any other value is true. This is not generally recommended; the if statement is easier to understand (for you or a reader of your code) if *condition* is a logical expression.
- If *condition* is true, *statement* is executed, but if *condition* is false, nothing happens.
- *condition* may be a vector or a matrix, in which case it is true only if *all* of its elements are nonzero. A single zero element in a vector or matrix renders it false.

```
if condition statement, end
```

Here are more examples of logical expressions involving relational operators, with their meanings in parentheses:

```
b^2 < 4*a*c (b² < 4ac)
x >= 0 (x ≥ 0)
a ~= 0 (a ≠ 0)
b^2 == 4*a*c (b² = 4ac)
```

$$b\text{\textasciicircum}2 < 4*a*c \quad (b^2 < 4ac)$$
$$x >= 0 \quad (x \geq 0)$$
$$a \sim= 0 \quad (a \neq 0)$$
$$b\text{\textasciicircum}2 == 4*a*c \quad (b^2 = 4ac)$$

Remember to use the double equal sign (==) when testing for equality:

```
if x == 0; disp( 'x equals zero'); end
```

Exercises

The following statements all assign logical expressions to the variable x. See if you can correctly determine the value of x in each case before checking your answer with MATLAB.

```
(a) x = 3 > 2
(b) x = 2 > 3
```

```
(c) x = -4 <= -3
(d) x = 1 < 1
(e) x = 2 ~= 2
(f) x = 3 == 3
(g) x = 0 < 0.5 < 1
```

Did you get item (f)? `3 == 3` is a logical expression that is true since 3 is undoubtedly equal to 3. The value 1 (for true) is therefore assigned to `x`. After executing these commands type the command `whos` to find that the variable `x` is in the class of *logical* variables.

What about (g)? As a *mathematical* inequality,

$$0 < 0.5 < 1$$

is undoubtedly true from a nonoperational point of view. However, as a MATLAB operational expression, the left-hand `<` is evaluated first, `0 < 0.5`, giving 1 (true). Then the right-hand operation is performed, `1 < 1`, giving 0 (false). Makes you think, doesn't it?

2.8.2 The `if-else` construct

If you enter the two lines

```
x = 2;
if x < 0 disp( 'neg' ), else disp( 'non-neg'), end
```

do you get the message `non-neg`? If you change the value of `x` to −1 and execute the `if` again, do you get the message `neg` this time? Finally, if you try

```
if 79 disp( 'true' ), else disp('false' ), end
```

do you get `true`? Try other values, including 0 and some negative values.

Most banks offer differential interest rates. Suppose the rate is 9% if the amount in your savings account is less than $5000, but 12% otherwise. The Random Bank goes one step further and gives you a random amount in your account to start with! Run the following program a few times:

```
bal = 10000 * rand;

if bal < 5000
   rate = 0.09;
else
   rate = 0.12;
end
```

```
newbal = bal + rate * bal;
disp( 'New balance after interest compounded is:' )
format bank
disp( newbal )
```

Display the values of `bal` and `rate` each time from the command line to check that MATLAB has chosen the correct interest rate.

The basic form of `if-else` for use in a program file is

```
if condition
    statementsA
else
    statementsB
end
```

Note that

- *statementsA* and *statementsB* represent one or more statements.
- If *condition* is true, *statementsA* are executed, but if *condition* is false, *statementsB* are executed. This is essentially how you force MATLAB to choose between two alternatives.
- `else` is optional.

2.8.3 The one-line `if-else` statement

The simplest general form of `if-else` for use on one line is

```
if condition; statementA; else; statementB; end
```

Note the following:

- Commas (or semicolons) are essential between the various clauses.
- `else` is optional.
- `end` is mandatory; without it, MATLAB will wait forever.

2.8.4 `elseif`

Suppose the Random Bank now offers 9% interest on balances of less than $5000, 12% for balances of $5000 or more but less than $10,000, and 15%

for balances of $10,000 or more. The following program calculates a customer's new balance after one year according to this scheme:

```
bal = 15000 * rand;

if bal < 5000
  rate = 0.09;
elseif bal < 10000
  rate = 0.12;
else
  rate = 0.15;
end

newbal = bal + rate * bal;
format bank
disp( 'New balance is:' )
disp( newbal )
```

Run the program a few times, and once again display the values of bal and rate each time to convince yourself that MATLAB has chosen the correct interest rate.

In general, the elseif clause is used:

```
if  condition1
    statementsA
elseif  condition2
    statementsB
elseif  condition3
    statementsC
...
else
    statementsE
end
```

This is sometimes called an elseif *ladder*. It works as follows:

1. *condition1* is tested. If it is true, *statementsA* are executed; MATLAB then moves to the next statement after end.
2. If *condition1* is false, MATLAB checks *condition2*. If it is true, *statementsB* are executed, followed by the statement after end.
3. In this way, all conditions are tested until a true one is found. As soon as a true condition is found, no further elseifs are examined and MATLAB jumps off the ladder.

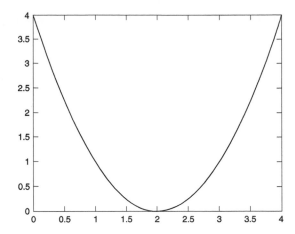

FIGURE 2.2
Quadratic function with equal roots.

4. If none of the conditions is true, *statements* after `else` are executed.
5. Arrange the logic so that not more than one of the conditions is true.
6. There can be any number of `elseif`s, but at most one `else`.
7. `elseif` *must* be written as one word.
8. It is good programming style to indent each group of statements as shown.

2.8.5 Logical operators

More complicated logical expressions can be constructed using the three *logical operators*: & (and), | (or), and ~ (not). For example, the quadratic equation

$$ax^2 + bx + c = 0$$

has equal roots, given by $-b/(2a)$, provided that $b^2 - 4ac = 0$ and $a \neq 0$ (Fig. 2.2). This translates into the following MATLAB statements:

```
if (b ^ 2 - 4*a*c == 0) & (a ~= 0)
      x = -b / (2*a);
end
```

Of course, a, b, and c must be assigned values prior to reaching this set of statements. Note the double equal sign in the test for equality; see Chapter 5 for more on logical operators.

2.8.6 Multiple `if`s versus `elseif`

You could have written the Random Bank program as follows:

```
bal = 15000 * rand;

if bal < 5000
    rate = 0.09;
end
if bal >= 5000 & bal < 10000
    rate = 0.12;
end
if bal >= 10000
    rate = 0.15;
end

newbal = bal + rate * bal;
format bank
disp( 'New balance is' )
disp( newbal )
```

However, this is inefficient since each of the three conditions is always tested, even if the first one is true. In the earlier `elseif` version, MATLAB jumps off the `elseif` ladder as soon as it finds a true condition. This saves a lot of computing time (and is easier to read) if the `if` construct is in a loop that is repeated often.

Using this form, instead of the `elseif` ladder, you can make the following common mistake:

```
if bal < 5000
    rate = 0.09;
end
if bal < 10000
    rate = 0.12;
end
if bal >= 10000
    rate = 0.15;
end
```

Can you see why you get the wrong answer (1120 instead of 1090) if `bal` has the value 1000? When designing the logic, you need to make sure that one and only one of the conditions will be true at any one time.

Another mistake frequently made is to replace the second `if` with something like

```
if 5000 < bal < 10000
    rate = 0.12;
end
```

which is compelling, as we saw above. However, *whatever* the value of bal, this condition will *always* be true. Can you see why? (Note that if bal is greater than 5000—for example, bal = 20000—the numerical truth value of the first test, namely, 5000 < bal, is true and hence has the numerical value of 1 since 1 is always less than 10000, even if bal = 20000.)

2.8.7 Nested ifs

An if construct can contain further ifs and so on. This is called *nesting* and should not be confused with the elseif ladder. You have to be careful with elses. In general, else belongs to the most recent if that has not been ended. The correct positioning of end is therefore very important, as the next example demonstrates.

Suppose you want to compute the solution to a quadratic equation. You may want to check whether $a = 0$ to prevent a division by zero. Your program could contain the following nested ifs:

```
...
d = b ^ 2 - 4*a*c;
if a ~= 0
  if d < 0
    disp( 'Complex roots' )
  else
    x1 = (-b + sqrt( d )) / (2*a);
    x2 = (-b - sqrt( d )) / (2*a);
  end    % first end  <<<<<<<<<<<<
end
```

The else belongs to the second if by default, as intended.

Now move the first end up as follows:

```
d = b ^ 2 {-} 4*a*c;
if a $^\sim$= 0
  if d < 0
    disp( 'Complex roots' )
  end    \% first end moved up now  <<<<<<<<<<<<
  else
    x1 = (-b + sqrt( d )) / (2*a);
    x2 = (-b {-} sqrt( d )) / (2*a);

end
```

The end that has been moved now closes the second if. The result is that else belongs to the *first* if instead of to the second one. Division by zero is therefore guaranteed instead of prevented!

2.8.8 Vectorizing ifs?

You may be wondering if for statements enclosing ifs can be vectorized. The answer is yes, courtesy of *logical arrays*. Discussion of this rather interesting topic is postponed until Chapter 5.

2.8.9 The switch statement

switch executes certain statements based on the value of a variable or expression. In this example it is used to decide whether a random integer is 1, 2, or 3 (see Section 5.1.5 for an explanation of this use of rand):

```
d = floor(3*rand) + 1
switch d
case 1
    disp( 'That''s a 1!' );
case 2
    disp( 'That''s a 2!' );
otherwise
    disp( 'Must be 3!' );
end
```

Multiple expressions can be handled in a single case statement by enclosing the case expression in a cell array (see Chapter 10):

```
d = floor(10*rand);
switch d
case {2, 4, 6, 8}
    disp( 'Even' );
case {1, 3, 5, 7, 9}
    disp( 'Odd' );
otherwise
    disp( 'Zero' );
end
```

2.9 COMPLEX NUMBERS

If you are not familiar with *complex* numbers, you can safely skip this section. However, it is useful to know what they are since the square root of a negative number may come up as a mistake if you are trying to work only with *real* numbers.

It is very easy to handle complex numbers in MATLAB. The special values i and j stand for $\sqrt{-1}$. Try sqrt(-1) to see how MATLAB represents complex numbers.

The symbol i may be used to assign complex values, for example,

```
z = 2 + 3*i
```

represents the complex number $2 + 3i$ (*real part 2, imaginary part 3*). You can also input a complex value like this:

```
2 + 3*i
```

in response to the input prompt (remember, no semicolon). The imaginary part of a complex number may also be entered without an asterisk, 3i.

All of the arithmetic operators (and most functions) work with complex numbers, such as sqrt(2 + 3*i) and exp(i*pi). Some functions are specific to complex numbers. If z is a complex number, real(z), imag(z), conj(z), and abs(z) all have the obvious meanings.

A complex number may be represented in polar coordinates:

$$z = re^{i\theta}$$

angle(z) returns θ between $-\pi$ and π; that is, atan2(imag(z), real(z)). abs(z) returns the magnitude r.

Since $e^{i\theta}$ gives the unit circle in polars, complex numbers provide a neat way of plotting a circle. Try the following:

```
circle = exp( 2*i*[1:360]*pi/360 );
plot(circle)
axis('equal')
```

Note these points:

- If y is complex, the statement plot(y) is equivalent to

  ```
  plot(real(y), imag(y))
  ```

- The statement axis('equal') is necessary to make circles look round; it changes what is known as the *aspect ratio* of the monitor. axis('normal') gives the default aspect ratio.

If you are using complex numbers, be careful not to use i or j for other variables; the new values will replace the value of $\sqrt{-1}$ and will cause nasty problems.

For complex matrices, the operations ' and .' behave differently. The ' operator is the *complex conjugate transpose*, meaning rows and columns are interchanged and signs of imaginary parts are changed. The .' operator, on the

other hand, does a pure transpose without taking the complex conjugate. To see this, set up a complex matrix `a` with the statement

```
a = [1+i 2+2i ; 3+3i 4+4i]
```

which results in

```
a =
   1.0000 + 1.0000i   2.0000 + 2.0000i
   3.0000 + 3.0000i   4.0000 + 4.0000i
```

The statement

```
        a'
```

then results in the complex conjugate transpose

```
ans =
   1.0000 - 1.0000i   3.0000 - 3.0000i
   2.0000 - 2.0000i   4.0000 - 4.0000i
```

whereas the statement

```
        a.'
```

results in the pure transpose

```
ans =
   1.0000 + 1.0000i   3.0000 + 3.0000i
   2.0000 + 2.0000i   4.0000 + 4.0000i
```

SUMMARY

- The MATLAB desktop consists of a number of tools: the Command Window, the Workspace browser, the Current Directory browser, and the Command History window.
- MATLAB has a comprehensive online Help system. It can be accessed through the Help button (?) on the desktop toolbar or the **Help** menu in any tool.
- A MATLAB program can be written in the Editor and cut and pasted to the Command Window (or it can be executed from the editor by clicking the green right arrow in the toolbar at the top of the Editor window).

- A script file is a text file (created by the MATLAB Editor or any other text editor) containing a collection of MATLAB statements. In other words, it is a program. The statements are carried out when the script file name is entered at the prompt in the Command Window. A script file name must have the `.m` extension. Script files are therefore also called M-files.
- The recommended way to run a script is from the Current Directory browser. The output from the script will then appear in the Command Window.
- A variable name consists only of letters, digits, and underscores, and must start with a letter. Only the first 63 characters are significant. MATLAB is case-sensitive by default. All variables created during a session remain in the workspace until removed with `clear`. The command `who` lists the variables in the workspace; `whos` gives their sizes.
- MATLAB refers to all variables as *arrays*, whether they are scalars (single-valued arrays), vectors, (1D arrays), or matrices (2D arrays).
- MATLAB names are case-sensitive.
- The Workspace browser on the desktop provides a handy visual representation of the workspace. Clicking a variable in it invokes the Array Editor, which may be used to view and change variable values.
- Vectors and matrices are entered in square brackets. Elements are separated by spaces or commas. Rows are separated by semicolons. The colon operator is used to generate vectors, with elements increasing (decreasing) by regular increments (decrements). Vectors are row vectors by default. Use the apostrophe transpose operator (`'`) to change a row vector into a column vector.
- An element of a vector is referred to by a subscript in parentheses. A subscript may itself be a vector. Subscripts always start at 1.
- The `diary` command copies everything that subsequently appears in the Command Window to the specified text file until the command `diary off` is given.
- Statements on the same line may be separated by commas or semicolons.
- A statement may be continued to the next line with an ellipsis of at least three dots.
- Numbers may be represented in fixed-point decimal notation or in floating-point scientific notation.
- MATLAB has 14 data types. The default numeric type is double precision. All mathematical operations are carried out in double precision.
- The six arithmetic operators for scalars are `+`, `-`, `*`, `\`, `/`, and `^`. They operate according to rules of precedence.
- An expression is a rule for evaluating a formula using numbers, operators, variables, and functions. A semicolon after an expression suppresses display of its value.
- Array operations are element by element between vectors or between scalars and vectors. The array operations of multiplication, right and left division, and exponentiation are indicated by `.*`, `./`, `.\`, and `.^` to

distinguish them from vector and matrix operations of the same name. They may be used to evaluate a formula repeatedly for some or all of the elements of a vector. This is called vectorization of the formula.

- `disp` is used to output (display) numbers and strings. `num2str` is useful with `disp` for displaying strings and numbers on the same line.
- The `format` command controls the way output is displayed.
- When vectors are displayed, a common scale factor is used if the elements are very large or very small, or differ greatly in magnitude.
- The `for` statement is used to repeat a group of statements a fixed number of times. If the index of a `for` statement is used in the expression being repeated, the expression can often be vectorized, saving a great deal of computing time.
- `tic` and `toc` may be used as a stopwatch.
- Logical expressions have the value true (1) or false (0) and are constructed with the six relational operators `>`, `>=`, `<`, `<=`, `==`, and `~=`. Any expression that evaluates to zero is regarded as false. Any other value is true. More complicated logical expressions can be formed from other logical expressions using the logical operators `&` (and), `|` (or), and `~` (not).
- `if-else` executes different groups of statements according to whether a logical expression is true or false. The `elseif` ladder is a good way to choose between a number of options, only one of which should be true at a time.
- `switch` enables choices to be made between discrete cases of a variable or expression.
- A string is a collection of characters enclosed in apostrophes.
- Complex numbers may be represented using the special variables `i` and `j`, which stand for the unit imaginary number $\sqrt{-1}$.

EXERCISES

2.1 Decide which of the following numbers are not acceptable in MATLAB, and state why:
 (a) `9,87`
 (b) `.0`
 (c) `25.82`
 (d) `-356231`
 (e) `3.57*e2`
 (f) `3.57e2.1`
 (g) `3.57e+2`
 (h) `3,57e-2`

2.2 State, giving reasons, which of the following are not valid MATLAB variable names:
 (a) `a2`
 (b) `a.2`

 (c) 2a

 (d) 'a'one

 (e) aone

 (f) _x_1

 (g) miXedUp

 (h) pay day

 (i) inf

 (j) Pay_Day

 (k) min*2

 (l) what

2.3 Translate the following expressions into MATLAB:

 (a) $p + \dfrac{w}{u}$

 (b) $p + \dfrac{w}{u+v}$

 (c) $\dfrac{p + \frac{w}{u+v}}{p + \frac{w}{u-v}}$

 (d) $x^{1/2}$

 (e) y^{y+z}

 (f) x^{y^z}

 (g) $(x^y)^z$

 (h) $x - x^3/3! + x^5/5!$

2.4 Translate the following into MATLAB statements:

 (a) Add 1 to the value of i and store the result in i.

 (b) Cube i, add j to this, and store the result in i.

 (c) Set g equal to the larger of the two variables e and f.

 (d) If d is greater than 0, set x equal to −b.

 (e) Divide the sum of a and b by the product of c and d, and store the result in x.

2.5 What's wrong with the following MATLAB statements?

 (a) n + 1 = n;

 (b) Fahrenheit temp = 9*C/5 + 32;

 (c) 2 = x;

2.6 Write a program to calculate x, where

$$x = \frac{-b + \sqrt{b^2 - 4ac}}{2a}$$

and $a = 2$, $b = -10$, $c = 12$. (Answer: 3.0)

2.7 There are eight pints in a gallon and 1.76 pints in a liter. The volume of a tank is given as 2 gallons and 4 pints. Write a script that inputs this volume in gallons and pints and converts it to liters. (Answer: 11.36)

2.8 Write a program to calculate gasoline consumption. It should assign the distance traveled (in kilometers) and the amount of gas used (in liters) and compute the consumption in km/liter as well as in the more usual

form of liters/100 km. Write some helpful headings so that your output looks something like this:

```
Distance     Liters used      km/L     L/100km
   528          46.23         11.42      8.76
```

2.9 Write some statements in MATLAB that exchange the contents of two variables a and b, using only one additional variable t.

2.10 Try Exercise 2.9 *without* using any additional variables!

2.11 If C and F are Celsius and Fahrenheit temperatures, respectively, the formula for conversion from Celsius to Fahrenheit is $F = 9C/5 + 32$.

 (a) Write a script that will ask you for the Celsius temperature and display the Fahrenheit equivalent with some sort of comment, such as

```
The Fahrenheit temperature is:...
```

 Try it out on the following Celsius temperatures (answers in parentheses): 0 (32), 100 (212), −40 (−40!), 37 (normal human temperature: 98.6).

 (b) Change the script to use vectors and array operations to compute the Fahrenheit equivalents of Celsius temperatures ranging from 20° to 30° in steps of 1°, and display them in two columns with a heading, like this:

```
Celsius    Fahrenheit
 20.00          68.00
 21.00          69.80
  ...
 30.00          86.00
```

2.12 Generate a table of conversions from degrees (first column) to radians (second column). Degrees should go from 0° to 360° in steps of 10°. Recall that π radians $= 180°$.

2.13 Set up a matrix (table) with degrees in the first column from 0 to 360 in steps of 30, sines in the second column, and cosines in the third column. Now try to add tangents in the fourth column. Can you figure out what's going on? Try some variations of the format command.

2.14 Write some statements that display a list of integers from 10 to 20 inclusive, each with its square root next to it.

2.15 Write a single statement to find and display the sum of the successive *even* integers 2, 4, ... , 200. (Answer: 10,100)

2.16 Ten students in a class take a test. The marks are out of 10. All the marks are entered in a MATLAB vector, marks. Write a statement to find and display the average mark. Try it on the following:

```
5  8  0  10  3  8  5  7  9 4  (Answer: 5.9)
```

Hint: Use the mean function.

2.17 What are the values of x and a after the following statements have been executed?

(a) `a = 0;`
(b) `i = 1;`
(c) `x = 0;`
(d) `a = a + i;`
(e) `x = x + i / a;`
(f) `a = a + i;`
(g) `x = x + i / a;`
(h) `a = a + i;`
(i) `x = x + i / a;`
(j) `a = a + i;`
(k) `x = x + i / a;`

2.18 Rewrite the statements in Exercise 2.17 more economically by using a `for` loop. Can you do even better by vectorizing the code?

2.19 Work out by hand the output of the following script for $n = 4$:

```
n = input( 'Number of terms? ' );
   s = 0;
for k = 1:n
   s = s + 1 / (k ^ 2);
end;
disp(sqrt(6 * s))
```

If you run this script for larger and larger values of n, you will find that the output approaches a well-known limit. Can you figure out what it is? Now rewrite the script using vectors and array operations.

2.20 Work through the following script by hand. Draw up a table of the values of i, j, and m to show how they change while the script executes. Check your answers by running the script.

```
v = [3 1 5];
   i = 1;

    for j = v
        i = i + 1;
        if i == 3
            i = i + 2;
            m = i + j;
        end
    end
```

2.21 The steady-state current I flowing in a circuit that contains a resistance $R = 5$, capacitance $C = 10$, and inductance $L = 4$ in series is given by

$$I = \frac{E}{\sqrt{R^2 + (2\pi\omega L - \frac{1}{2\pi\omega C})^2}}$$

where $E = 2$ and $\omega = 2$ are the input voltage and angular frequency, respectively. Compute the value of I. (Answer: 0.0396)

2.22 The electricity accounts of residents in a very small town are calculated as follows:

- If 500 units or fewer are used, the cost is 2 cents per unit.
- If more than 500 but not more than 1000 units are used, the cost is $10 for the first 500 units and 5 cents for every unit in excess of 500.
- If more than 1000 units are used, the cost is $35 for the first 1000 units plus 10 cents for every unit in excess of 1000.
- A basic service fee of $5 is charged, no matter how much electricity is used.

Write a program that enters the following five consumptions into a vector and uses a for loop to calculate and display the total charge for each one: 200, 500, 700, 1000, 1500. (Answers: $9, $15, $25, $40, $90)

2.23 Suppose you deposit $50 in a bank account every month for a year. Every month, after the deposit has been made, interest at the rate of 1% is added to the balance: After one month the balance is $50.50, and after two months it is $101.51. Write a program to compute and print the balance each month for a year. Arrange the output to look something like this:

```
MONTH          MONTH-END BALANCE

  1                   50.50
  2                  101.51
  3                  153.02
 ...
 12                  640.47
```

2.24 If you invest $1000 for one year at an interest rate of 12%, the return is $1120 at the end of the year. But if interest is compounded at the rate of 1% *monthly* (i.e., 1/12 of the annual rate), you get slightly more interest because it is compounded. Write a program that uses a for loop to compute the balance after a year of compounding interest in this way. The answer should be $1126.83. Evaluate the formula for this result separately as a check: 1000×1.01^{12}.

2.25 A plumber opens a savings account with $100,000 at the beginning of January. He then makes a deposit of $1000 at the end of each month for the next 12 months (starting at the end of January). Interest is calculated and added to his account at the end of each month (before the $1000 deposit is made). The monthly interest rate depends on the amount A in his account at the time interest is calculated, in the following way:

$$A \le 1\ 10\ 000: \qquad 1\%$$
$$1\ 10\ 000 < A \le 1\ 25\ 000: \quad 1.5\%$$
$$A > 1\ 25\ 000: \qquad 2\%$$

Write a program that displays, under suitable headings, for each of the 12 months, the situation at the end of the month as follows: the number of the month, the interest rate, the amount of interest, and the new balance. (Answer: Values in the last row of output should be 12, 0.02, 2534.58, 130263.78).

2.26 It has been suggested that the population of the United States may be modeled by the formula

$$P(t) = \frac{19\ 72\ 73\ 000}{1 + e^{-0.03134(t - 1913.25)}}$$

where t is the date in years. Write a program to compute and display the population every `ten` years from 1790 to 2000. Try to plot a graph of the population against time as well. Use your program to find out if the population ever reaches a "steady state" (i.e., stops changing).

2.27 A mortgage bond (loan) of amount L is obtained to buy a house. The interest rate r is 15%. The fixed monthly payment P that will pay off the bond loan over N years is given by the formula

$$P = \frac{rL(1 + r/12)^{12N}}{12[(1 + r/12)^{12N} - 1]}$$

(a) Write a program to compute and print P if $N = 20$ and the bond is for $50,000. You should get $658.39.

(b) See how P changes with N by running the program for different values of N (use `input`). Can you find a value for which the payment is less than $625?

(c) Go back to $N = 20$ and examine the effect of different interest rates. You should see that raising the interest rate by 1% (0.01) increases the monthly payment by about $37.

2.28 It is useful to work out how the period of a bond repayment changes if you increase or decrease P. The formula for N is given by

$$N = \frac{\ln\left(\frac{P}{P - rL/12}\right)}{12 \ln(1 + r/12)}$$

(a) Write a new program to compute this formula. Use the built-in function `log` for the natural logarithm ln. How long will it take to pay off a loan of $50,000 at $800 a month if the interest remains at 15%? (Answer: 10.2 years—nearly twice as fast as when paying $658 a month.)

(b) Use your program to find out by trial and error the smallest monthly payment that will pay off the loan this side of eternity. *Hint*: recall that it is not possible to find the logarithm of a negative number, so P must not be less than $rL/12$.

APPENDIX 2.A SUPPLEMENTARY MATERIAL

Supplementary material related to this chapter can be found online at https://doi.org/10.1016/B978-0-08-102997-8.00008-7.

Program Design and Algorithm Development

THE OBJECTIVES OF THIS CHAPTER ARE TO INTRODUCE YOU TO:

- Program design
- User-defined MATLAB functions

CONTENTS

The program design process . 84
The projectile problem 87
Programming MATLAB functions 92
Inline objects: Harmonic oscillators 92
MATLAB function: $y = f(x)$ 94
Summary 96
Exercises 97

This chapter is an introduction to the design of computer programs. The top-down design process is elaborated to help you think about good problem-solving strategies as they relate to the design of procedures for using software like MATLAB. We will consider the design of your own toolbox to be included among those already available with your version of MATLAB, such as Simulink, Symbolic Math, and Controls System. This is a big advantage of MATLAB (and tools like it); it allows you to customize your working environment to meet your own needs. It is not only the "mathematics handbook" of today's student, engineer, and scientist, it is also a useful environment to develop software tools that go beyond any handbook to help you to solve relatively complicated mathematical problems.

In the first part of this chapter we discuss the design process. In the second part we examine the structure plan—the detailed description of the algorithm to be implemented. We will consider relatively simple programs. However, the process described is intended to provide insight into what you will confront when you deal with more complex engineering, scientific, and mathematical problems during the later years of your formal education, your life-long learning, and your continuing professional education. In the third part we introduce the basic construct of a MATLAB function to help you develop more sophisticated programs.

To be sure, the examples examined so far have been logically simple. This is because we have been concentrating on the technical aspects of writing correct MATLAB statements. It is very important to learn how MATLAB does the arithmetic operations that form the basis of more complex programs. To design a successful program you need to understand a problem thoroughly and break it down into its most fundamental logical stages. In other words, you have to develop a systematic procedure or *algorithm* for solving it.

83

Essential MATLAB for Engineers and Scientists. https://doi.org/10.1016/B978-0-08-102997-8.00009-9

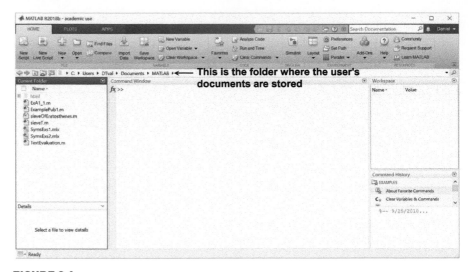

FIGURE 3.1
Creating your work folder: In your Users-Documents folder.

There are a number of methods that may assist in algorithm development. In this chapter we look at one, the *structure plan*. Its development is the primary part of the software (or code) design process because it is the steps in it that are translated into a language the computer can understand—for example, into MATLAB commands.

3.1 THE PROGRAM DESIGN PROCESS

Useful utilities translated into MATLAB (either sequences of command lines or *functions*, which are described later in the text) and saved as M-files in your working directory are your primary goals (see Fig. 3.1). There are numerous toolboxes available through MathWorks (among others) on a variety of engineering and scientific topics. A great example is the Aerospace Toolbox, which provides reference standards, environmental models, and aerodynamic coefficients importing for advanced aerospace engineering designs. Explore the MathWorks Web site for products available (http://www.mathworks.com/).

In your working directory, (e.g., the folder we are currently in, viz., \MATLAB), you will begin to accumulate a set of M-files that you have created as you use MATLAB.

Certainly, you want to be sure that the tools you save are reasonably well written (i.e., reasonably well designed). What does it mean to create well-written programs?

The goals in designing a software tool are that it works, it can easily be read and understood, and, hence, it can be systematically modified when required.

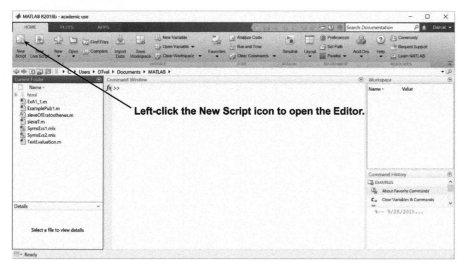

FIGURE 3.2
Creating your work folder.

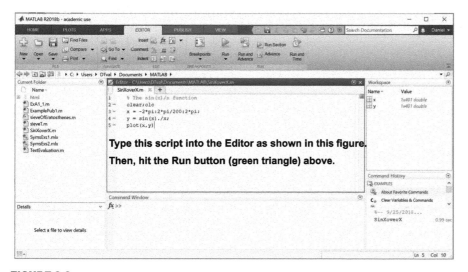

FIGURE 3.3
Creating your work folder: Entering, saving and executing a MATLAB M-file.

For programs to work well they must satisfy the requirements associated with the problem or class of problems they are intended to solve. The specifications (i.e., the detailed description of purpose, or function, inputs, method of processing, outputs, and any other special requirements) must be known to design an effective algorithm or computer program, which must work completely and correctly. That is, all options should be usable without error within the limits of the specifications (see Figs. 3.2 and 3.3).

The program must be readable and hence clearly understandable. Thus, it is useful to decompose major tasks (or the main program) into subtasks (or sub-programs) that do specific parts of it. It is much easier to read subprograms, which have fewer lines, than one large main program that doesn't segregate the subtasks effectively, particularly if the problem to be solved is relatively complicated. Each subtask should be designed so that it can be evaluated independently before it is implemented in the larger scheme of things (i.e., in the main program plan).

A well written code, when it works, is much more easily evaluated in the testing phase of the design process. If changes are necessary to correct sign mistakes and the like, they can be easily implemented. One thing to keep in mind when you add comments to describe the process programmed is this: Add enough comments and references so that a year from the time you write the program you know exactly what was done and for what purpose. Note that the first few comment lines in a script file are displayed in the Command Window when you type `help` followed by the name of your file (file naming is also an art).

The design process[1] is outlined next. The steps may be listed as follows:

Step 1 *Problem analysis.* The context of the proposed investigation must be established to provide the proper motivation for the design of a computer program. The designer must fully recognize the need and must develop an understanding of the nature of the problem to be solved.

Step 2 *Problem statement.* Develop a detailed statement of the mathematical problem to be solved with a computer program.

Step 3 *Processing scheme.* Define the inputs required and the outputs to be produced by the program.

Step 4 *Algorithm.* Design the step-by-step procedure in a *top-down* process that decomposes the overall problem into subordinate problems. The subtasks to solve the latter are refined by designing an itemized list of steps to be programmed. This list of tasks is the *structure plan* and is written in *pseudo-code* (i.e., a combination of English, mathematics, and anticipated MATLAB commands). The goal is a plan that is understandable and easily translated into a computer language.

Step 5 *Program algorithm.* Translate or convert the algorithm into a computer language (e.g., MATLAB) and debug the syntax errors until the tool executes successfully.

Step 6 *Evaluation.* Test all of the options and conduct a validation study of the program. For example, compare results with other programs that do similar tasks, compare with experimental data if appropriate, and compare with theoretical predictions based on theoretical methodology related to

[1] For a more detailed description of software design technology see, for example, *C++ Data Structures* by Nell Dale (Jones and Bartlett, 1998).

the problems to be solved. The objective is to determine that the sub-tasks and the overall program are correct and accurate. The additional debugging in this step is to find and correct *logical* errors (e.g., mistyping of expressions by putting a plus sign where a minus sign was supposed to be) and *runtime* errors that may occur after the program successfully executes (e.g., cases where division by zero unintendedly occurs).

Step 7 *Application.* Solve the problems the program was designed to solve. If the program is well designed and useful, it can be saved in your working directory (i.e., in your user-developed toolbox) for future use.

3.1.1 The projectile problem

Step 1. Let us consider the projectile problem examined in first-semester physics. It is assumed that engineering and science students understand this problem (if it is not familiar to you, find a physics text that describes it or search the Web; the formulas that apply will be provided in step 2).

In this example we want to calculate the flight of a projectile (e.g., a golf ball) launched at a prescribed speed and a prescribed launch angle. We want to determine the trajectory of the flight path and the horizontal distance the projectile (or object) travels before it hits the ground. Let us assume zero air resistance and a constant gravitational force acting on the object in the opposite direction of the vertical distance from the ground. The launch angle, θ_o, is defined as the angle measured from the horizontal (ground plane) upward toward the vertical direction, $0 < \theta_o \leq \pi/2$, where $\theta_o = 0$ implies a launch in the horizontal direction and $\theta_o = \pi/2$ implies a launch in the vertical direction (i.e., in the opposite direction of gravity). If $g = 9.81$ m/s^2 is used as the acceleration of gravity, the launch speed, V_o, must be entered in units of m/s. Thus, if the time, $t > 0$, is the time in seconds (s) from the launch time of $t = 0$, the distance traveled in x (the horizontal direction) and y (the vertical direction) is in meters (m).

We want to determine the time it takes the projectile, from the start of motion, to hit the ground, the horizontal distance traveled, and the shape of the trajectory. In addition, we want to plot the speed of the projectile versus the angular direction of this vector. We need, of course, the theory (or mathematical expressions) that describes the solution to the zero-resistance projectile problem in order to develop an algorithm to obtain solutions to it.

Step 2. The mathematical formulas that describe the solution to the projectile problem are provided in this step. Given the launch angle and launch speed, the horizontal distance traveled from $x = y = 0$, which is the coordinate location of the launcher, is

$$x_{max} = 2\frac{V_o^2}{g}\sin\theta_o\cos\theta_o$$

The time from $t = 0$ at launch for the projectile to reach x_{max} (i.e., its range) is

$$t_{x_{max}} = 2\frac{V_o}{g}\sin\theta_o$$

The object reaches its maximum altitude,

$$y_{max} = \frac{V_o^2}{2g}\sin^2\theta_o$$

at time

$$t_{y_{max}} = \frac{V_o}{g}\sin\theta_o$$

The horizontal distance traveled when the object reaches the maximum altitude is $x_{y_{max}} = x_{max}/2.0$.

The trajectory (or flight path) is described by the following pair of coordinates at a given instant of time between $t = 0$ and $t_{x_{max}}$:

$$x = V_o t \cos\theta_o$$
$$y = V_o t \sin\theta_o - \frac{g}{2}t^2$$

We need to solve these equations over the range of time $0 < t \le t_{x_{max}}$ for prescribed launch conditions $V_o > 0$ and $0 < \theta \le \pi/2$. Then the maximum values of the altitude and the range are computed along with their respective arrival times. Finally, we want to plot V versus θ, where

$$V = \sqrt{(V_o \cos\theta_o)^2 + (V_o \sin\theta_o - gt)^2}$$

and

$$\theta = \tan^{-1}\left(\frac{V_y}{V_x}\right)$$

We must keep in mind when we study the solutions based on these formulas that the air resistance was assumed negligible and the gravitational acceleration was assumed constant.

Step 3. The required inputs are g, V_o, θ_o, and a finite number of time steps between $t = 0$ and the time the object returns to the ground. The outputs are the range and time of flight, the maximum altitude and the time it is reached, and the shape of the trajectory in graphical form.

Steps 4 and 5. The algorithm and structure plan developed to solve this problem are given next as a MATLAB program, because it is relatively straightforward and the translation to MATLAB is well commented with details of the approach applied to its solution (i.e., the steps of the structure plan are enumerated). This plan, and M-file, of course, is the summary of the results developed by trying a number of approaches during the design process, and thus discarding numerous sheets of scratch paper before summarizing the results! (There are more explicit examples of structure plans for your review and investigation in the next section of this chapter.) Keep in mind that it was not difficult to enumerate a list of steps associated with the general design process, that is, the technical problem solving. However, it is certainly not so easy to implement the steps because they draw heavily on your technical-solution design experience. Hence, we must begin by studying the design of relatively simple programs like the one described in this section.

The evaluated and tested code is as follows:

```
% The proctile problem with zero air resistance
% in a gravitational field with constant g.
% Written by Daniel T. Valentine .. September 2006
% Revised by D. T. Valentine ...... 2012/2016/2018
% An eight-step structure plan applied in MATLAB:
%
% 1. Define the input variables.
%
g = 9.81; % Gravity in m/s/s.
vo = input('What is the launch speed in m/s? ')
tho = input('What is the launch angle in degrees? ')
tho = pi*tho/180; % Conversion of degrees to radians.
%
% 2. Calculate the range and duration of the flight.
%
txmax = (2*vo/g) * sin(tho);
xmax = txmax * vo * cos(tho);
%
% 3. Calculate the sequence of time steps to compute
% trajectory.
%
dt = txmax/100;
t = 0:dt:txmax;
%
% 4. Compute the trajectory.
%
x = (vo * cos(tho)) .* t;
y = (vo * sin(tho)) .* t - (g/2) .* t.^2;
```

```
%
% 5. Compute the speed and angular direction of the
% projectile. Note that vx = dx/dt, vy = dy/dt.
%
vx = vo * cos(tho);
vy = vo * sin(tho) - g .* t;
v = sqrt(vx.*vx + vy.*vy);      % Speed
th = (180/pi) .* atan2(vy,vx); % Angular direction
%
% 6. Compute the time and horizontal distance at the
% maximum altitude.
%
tymax = (vo/g) * sin(tho);
xymax = xmax/2;
ymax = (vo/2) * tymax * sin(tho);
%
% 7. Display in the Command Window and on figures the ouput.
%
disp(['Range in meters = ',num2str(xmax),',' ...
    ' Duration in seconds = ', num2str(txmax)])
disp(' ')
disp(['Maximum altitude in meters = ',num2str(ymax), ...
    ', Arrival at this altitude in seconds = ', num2str(tymax)])
plot(x,y,'k',xmax,y(size(t)),'o',xmax/2,ymax,'o')
title(['Projectile flight path: v_o = ', num2str(vo),' m/s' ...
    ', \theta_o = ', num2str(180*tho/pi),' degrees'])
xlabel('x'), ylabel('y')      % Plot of Figure 1.
figure % Creates a new figure.
plot(v,th,'r')
title('Projectile speed vs. angle')
xlabel('V'), ylabel('\theta') % Plot of Figure 2.
%
% 8. Stop.
```

Steps 6 and 7. The program was evaluated by executing a number of values of the launch angle and launch speed within the required specifications. The angle of 45 degrees was checked to determine that the maximum range occurred at this angle for all specified launch speeds. This is well known for the zero air resistance case in a constant g force field. Executing this code for $V_o = 10$ m/s and $\theta_o = 45$ degrees, the trajectory and the plot of orientation versus speed in Figs. 3.4 and 3.5, respectively, were produced.

How can you find additional examples of MATLAB programs (good ones or otherwise) to help develop tools to solve your own problems? We all recognize that examples aren't a bad way of learning to use tools. New tools are

FIGURE 3.4
Projectile trajectory.

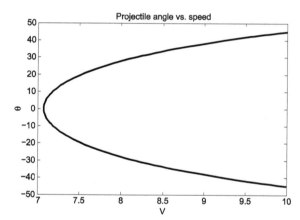

FIGURE 3.5
Projectile angle versus speed.

continually being developed by the users of MATLAB. If one proves to be of more general use, MathWorks may include it in their list of products (if, of course, the tools' author desires this). There are also many examples of useful scripts that are placed on the Web for anyone interested in them. They, of course, must be evaluated carefully since it is the user's responsibility, *not* the creator's, to ensure the correctness of their results. This responsibility holds for all tools applied by the engineer and the scientist. Hence, it is very important (just as in using a laboratory apparatus) that users prove to themselves that the tool they are using is indeed valid for the problem they are trying to solve.

To illustrate how easy it is to find examples of scripts, the author typed `MATLAB examples` in one of the available search engines and found the following (among many others):

```
t = 0:.1:2*pi;
subplot(2,2,1)
plot(t,sin(t))
subplot(2,2,2)
plot(t,cos(t))
subplot(2,2,3)
plot(t,exp(t))
subplot(2,2,4)
plot(t,1./(1+t.^2))
```

This script illustrates how to put four plots in a single figure window. To check that it works, type each line in the Command Window followed by **Enter**. Note the position of each graphic; location is determined by the three integers in the subplot function list of arguments. Search Help via the question mark (?) for more information on subplot.

3.2 PROGRAMMING MATLAB FUNCTIONS

MATLAB enables you to create your own function M-files. A function M-file is similar to a script file in that it also has a .m extension. However, a function M-file differs from a script file in that a function M-file communicates with the MATLAB workspace only through specially designated *input* and *output arguments*. Functions are indispensable tools when it comes to breaking a problem down into manageable logical pieces.

Short mathematical functions may be written as one-line *inline objects*. This capability is illustrated by example in the next subsection. In the subsequent subsection essential ideas embodied in the creation of a MATLAB function M-file will be introduced. (Further details on writing functions are provided in Chapter 10.)

3.2.1 Inline objects: Harmonic oscillators

If two coupled harmonic oscillators, e.g., two masses connected with a spring on a very smooth table, are considered as a single system, the output of the system as a function of time t could be given by something like

$$h(t) = \cos(8t) + \cos(9t). \tag{3.1}$$

You can represent $h(t)$ at the command line by creating an *inline object* as follows:

```
h = inline( 'cos(8*t) + cos(9*t)' );
```

Now write some MATLAB statements in the Command Window which use your function h to draw the graph in Fig. 3.6 e.g.,

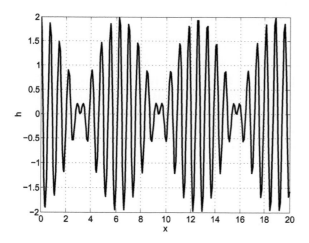

FIGURE 3.6
$\cos(8t) + \cos(9t)$.

```
x = 0 : 20/300 : 20;
plot(x, h(x)), grid
```

Note:

- The variable t in the `inline` definition of h is the *input argument*. It is essentially a 'dummy' variable, and serves only to provide input to the function from the outside world. You can use any variable name here; it doesn't have to be the same as the one used when you invoke (use) the function.
- You can create functions of more than one argument with `inline`. For example,

```
f1 = inline( 'x.^2 + y.^2', 'x', 'y' );
f1(1, 2)
ans =
     5
```

Note that the input values of x and y can be arrays and, hence, the output f1 will be an array of the same size as x and y. For example, after executing the previous two command lines to get the ans = 5, try the following:

```
x = [1 2 3; 1 2 3]; y = [4 5 6; 4 5 6];
f1(x,y)
ans =
    17    29    45
    17    29    45
```

The answer is an element-by-element operation that leads to the value of the function for each combination of like-located array elements.

3.2.2 MATLAB function: $y = f(x)$

The essential features of a MATLAB function are embodied in the following example. Let us consider the function y defined by the relationship $y = f(x)$, i.e., y is a function of x. The construction of a function file starts with the declaration of a function command. This is followed by the formula that is the function of interest that you wish to substitute a particular value of x to get the corresponding value of y. The structure plan for the evaluation of a particular algebraic function is as follows:

```
1. function y = f(x) % where $x$ is the input and $y$ is the output.
2. y =   x.^3 - 0.95*x;
3. end % This is not necessary to include; however, it plays the
      role of STOP.
```

The function M-file created based on this plan is as follows;

```
function y = f(x) % where $x$ is the input and $y$ is the output.
y =   x.^3 - 0.95*x;
end % This is not necessary to include; however, it plays the role
      of STOP.
```

It was saved as `f.m`. After saving it, it was used as follows in the Command Window. Try the following example:

```
>> f(2)
ans =
    6.1000
```

Let us next create a function that takes as input the three coefficients of the quadratic equation, viz.,

$$ax^2 + bx + c = 0.$$

We want to determine the two roots of this equation, i.e., we want to determine the solution to this equation. One way of dealing with this problem is to apply the known solutions of the quadratic equation. Let us do this by creating a function file by following the structure plan for the complete algorithm for finding the solution(s) x, given the coefficients a, b and c given as follows:

QUADRATIC EQUATION STRUCTURE PLAN

1. Start
2. Input data (a, b, c)

3. If $a = 0$ then
 If $b = 0$ then
 If $c = 0$ then
 Display 'Solution indeterminate'
 else
 Display 'There is no solution'
 else
 $x = -c/b$
 Display x (only one root: equation is linear)
 else if $b^2 < 4ac$ then
 Display 'Complex roots'
 else if $b^2 = 4ac$ then
 $x = -b/(2a)$
 Display x (equal roots)
 else
 $x_1 = (-b + \sqrt{b^2 - 4ac})/(2a)$
 $x_2 = (-b - \sqrt{b^2 - 4ac})/(2a)$
 Display x_1, x_2
4. Stop.

A function file based on this plan is as follows:

```
function x = quadratic(a,b,c)
% Equation:
%     a*x^2 + b*x + c = 0
% Input: a,b,c
% Output: x = [x1 x2], the two solutions of
% this eequation.
if a==0 & b==0 & c==0
    disp(' ')
    disp('Solution indeterminate')
elseif a==0 & b==0
    disp(' ')
    disp('There is no solution')
elseif a==0
    disp(' ')
    disp('Only one root: equation is linear')
    disp('      x      ')
    x1 = -c/b;
    x2 = NaN;
elseif b^2 < 4*a*c
    disp(' ')
    disp(' x1, x2 are complex roots ')
    disp('           x1                  x2')
```

```
    x1 = (-b + sqrt(b^2 - 4*a*c))/(2*a);
    x2 = (-b - sqrt(b^2 - 4*a*c))/(2*a);
elseif b^2 == 4*a*c
    x1 = -b/(2*a);
    x2 = x1;
    disp('equal roots')
    disp('    x1    x2')
else
    x1 = (-b + sqrt(b^2 - 4*a*c))/(2*a);
    x2 = (-b - sqrt(b^2 - 4*a*c))/(2*a);
    disp('    x1    x2')
end
if a==0 & b==0 & c==0
elseif a==0 & b==0
else
    disp([x1 x2]);
end
end
```

This function is saved with the file name quadratic.m.

```
>> a = 4; b = 2; c = -2;
>> quadratic(a,b,c)
    x1    x2
    0.5000  -1.0000
>> who
Your variables are:
a  b  c
```

The two roots in this case are real. It is a useful exercise to test all possibilities to evaluate whether or not this function successfully deals with all quadratic equations with constant coefficients. The purpose of this example was to il-lustrate how to construct a function file. Note that the only variables in the **Workspace** are the coefficients a b c. The variables defined and needed in the function are not included in the Workspace. Hence, functions do not clutter the Workspace with the variables only needed in the function itself to execute the commands within it.

SUMMARY

- An algorithm is a systematic logical procedure for solving a problem.
- A structure plan is a representation of an algorithm in pseudo-code.
- A function M-file is a script file designed to handle a particular task that may be activated (invoked) whenever needed.

EXERCISES

The problems in these exercises should all be structure-planned before being written up as MATLAB programs (where appropriate).

3.1 The structure plan in this example defines a geometric construction. Carry out the plan by sketching the construction:
1. Draw two perpendicular x- and y-axes
2. Draw the points A (10, 0) and B (0, 1)
3. While A does not coincide with the origin repeat: Draw a straight line joining A and B Move A one unit to the left along the x-axis Move B one unit up on the y-axis
4. Stop

3.2 Consider the following structure plan, where M and N represent MATLAB variables:
1. Set M = 44 and N = 28
2. While M not equal to N repeat: While $M > N$ repeat: Replace value of M by $M - N$ While $N > M$ repeat: Replace value of N by $N - M$
3. Display M
4. Stop
 (a) Work through the structure plan, sketching the contents of M and N during execution. Give the output.
 (b) Repeat (a) for $M = 14$ and $N = 24$.
 (c) What general arithmetic procedure does the algorithm carry out (try more values of M and N if necessary)?

3.3 Write a program to convert a Fahrenheit temperature to Celsius. Test it on the data in Exercise 2.11 (where the reverse conversion is done).

3.4 Write a script that inputs any two numbers (which may be equal) and displays the larger one with a suitable message or, if they are equal, displays a message to that effect.

3.5 Write a script for the general solution to the quadratic equation $ax^2 + bx + c = 0$. Use the structure plan in Section 3.2.2. Your script should be able to handle all possible values of the data a, b, and c. Try it out on the following values:
 (a) 1, 1, 1 (complex roots)
 (b) 2, 4, 2 (equal roots of -1.0)
 (c) 2, 2, -12 (roots of 2.0 and -3.0)
The structure plan in Section 3.2.2 is for programming languages that cannot handle complex numbers. MATLAB can. Adjust your script so that it can also find complex roots. Test it on case (a); the roots are $-0.5 \pm 0.866i$.

3.6 Develop a structure plan for the solution to two simultaneous linear equations (i.e., the equations of two straight lines). Your algorithm must be able to handle all possible situations; that is, lines intersecting,

parallel, or coincident. Write a program to implement your algorithm, and test it on some equations for which you know the solutions, such as

$$x + y = 3$$
$$2x - y = 3$$

$(x = 2, y = 1)$.

Hint: Begin by deriving an algebraic formula for the solution to the system:

$$ax + by = c$$
$$dx + ey = f$$

The program should input the coefficients a, b, c, d, e, and f.

We will see in Chapter 16 that MATLAB has a very elegant way of solving systems of equations directly, using matrix algebra. However, it is good for the development of your programming skills to do it the long way, as in this exercise.

3.7 We wish to examine the motion of a damped harmonic oscillator. The small amplitude oscillation of a unit mass attached to a spring is given by the formula $y = e^{-(R/2)t} \sin(\omega_1 t)$, where $\omega_1^2 = \omega_o^2 - R^2/4$ is the square of the natural frequency of the oscillation with damping (i.e., with resistance to motion); $\omega_o^2 = k$ is the square of the natural frequency of undamped oscillation; k is the spring constant; and R is the damping coefficient. Consider $k = 1$ and vary R from 0 to 2 in increments of 0.5. Plot y versus t for t from 0 to 10 in increments of 0.1.

Hint: Develop a solution procedure by working backwards through the problem statement. Starting at the end of the problem statement, the solution procedure requires the programmer to assign the input variables first followed by the execution of the formula for the amplitude and ending with the output in graphical form.

3.8 Let's examine the shape of a uniform cable hanging under its own weight. The shape is described by the formula $y = \cosh(x/c)$. This shape is called a uniform catenary. The parameter c is the vertical distance from $y = 0$ where the bottom of the catenary is located. Plot the shape of the catenary between $x = -10$ and $x = 10$ for $c = 5$. Compare this with the same result for $c = 4$.

Hint: The hyperbolic cosine, cosh, is a built-in MATLAB function that is used in a similar way to the sine function, sin.

MATLAB Functions and Data Import-Export Utilities

THE OBJECTIVES OF THIS CHAPTER ARE:

(1) To enable you to become familiar with some of the more common MATLAB functions.
(2) To introduce you briefly to ways of importing and exporting data in and out of the MATLAB workspace using:
- The `load` and `save` commands.
- The Import Wizard.
- The low-level file input/output (I/O) functions.

CONTENTS

Common functions 99

Importing and exporting data .. 104
The load and save commands 104
Exporting text (ASCII) data 104
Importing text (ASCII) data 105
Exporting binary data 105
Importing binary data 106

Summary 106

Exercises 106

At this point you should be able to write a MATLAB program which inputs data, performs simple arithmetic operations on the data, perhaps involving loops and decisions, and displays the results of the computation in a comprehensible form. However, more interesting problems in science and engineering are likely to involve special mathematical functions like sines, cosines, logarithms, etc. MATLAB comes with a large collection of such functions; we have seen some of them already. This chapter introduces you to the more common functions available in MATLAB. In addition, you may wish to import data to be plotted or operated on mathematically, and export data for future use. Hence, this chapter also introduces you to the importing of data into the MATLAB workspace from various sources. It also discusses the exporting of data to files in your `work` directory for later use with MATLAB or with other software tools.

4.1 COMMON FUNCTIONS

Tables of MATLAB functions and commands appear in Appendix C. A short list of some of the more common ones follows. Use `helpwin` at the command line to see a list of categories of functions, with links to descriptions of the functions. Alternatively, go to the **Contents** listing in the Help Navigator (the left pane in the Help browser), and expand successively MATLAB, **Reference**, **MATLAB Function Reference** where you can choose either **Functions by Category**, or **Alphabetical List of Functions**.

Note that if the argument of a function is an array, the function is applied element by element to all the values in the array, e.g.,

Essential MATLAB for Engineers and Scientists. https://doi.org/10.1016/B978-0-08-102997-8.00010-5

```
sqrt([1 2 3 4])
```

returns

```
1.0000    1.4142    1.7321    2.0000
```

Since this text is written in a tutorial style, it is expected that you will examine the following list of some of the more common functions in MATLAB. It is also, to some extent, expected that you already know, from your first courses in mathematics and science, something about these functions. One way to examine these functions is to plot them. Have some fun experimenting with MATLAB in your investigation of the following functions! For example, do the following exercise for all functions of the assigned variable x, as illustrated next.

```
x = -1:.1:1; <Enter>
plot(x,abs(x),'o') <Enter>
```

You should get an illustration that looks like a V.

```
abs(x)
```
 absolute value of x.
```
acos(x)
```
 arc cosine (inverse cosine) of x between 0 and π.
```
acosh(x)
```
 inverse hyperbolic cosine of x, i.e., $\ln(x + \sqrt{x^2 - 1})$.
```
asin(x)
```
 arc sine (inverse sine) of x between $-\pi/2$ and $\pi/2$.
```
asinh(x)
```
 inverse hyperbolic sine of x, i.e., $\ln(x + \sqrt{x^2 + 1})$.
```
atan(x)
```
 arc tangent of x between $-\pi/2$ and $\pi/2$.
```
atan2(y, x)
```
 arc tangent of y/x between $-\pi$ and π.
```
atanh(x)
```
 inverse hyperbolic tangent of x, i.e., $\dfrac{1}{2}\ln\left(\dfrac{1+x}{1-x}\right)$.
```
ceil(x)
```
 smallest integer which exceeds x, i.e., rounds up to nearest integer, e.g., `ceil(-3.9)` returns -3, `ceil(3.9)` returns 4.
```
clock
```
 time and date in a six-element vector, e.g., the statements

```
t = clock;
fprintf( '  %02.0f:%02.0f:%02.0f\n', t(4), t(5), t(6) );
```

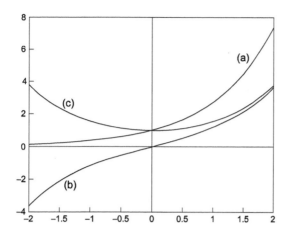

FIGURE 4.1
(a) Exponential, (b) hyperbolic sine and (c) hyperbolic cosine functions.

result in `14:09:03`. Note how the hours, minutes and seconds are left-filled with zeros if necessary.

`cos(x)`

cosine of x.

`cosh(x)`

hyperbolic cosine of x, i.e., $\dfrac{e^x + e^{-x}}{2}$ (see Fig. 4.1).

`cot(x)`

cotangent of x.

`csc(x)`

cosecant of x.

`cumsum(x)`

cumulative sum of the elements of x, e.g., `cumsum(1:4)` returns [1 3 6 10].

`date`

date in a string in dd-mmm-yyyy format, e.g., `02-Feb-2001`.

`exp(x)`

value of the exponential function e^x (see Fig. 4.1).

`fix(x)`

rounds to the nearest integer towards zero, e.g., `fix(-3.9)` returns `-3`, `fix(3.9)` returns 3.

`floor(x)`

largest integer not exceeding x, i.e., rounds down to nearest integer, e.g., `floor(-3.9)` returns `-4`, `floor(3.9)` returns 3.

`length(x)`

number of elements of vector x.

`log(x)`

natural logarithm of x.

```
log10(x)
```
 base 10 logarithm of x.

```
max(x)
```
 maximum element of vector x.

```
mean(x)
```
 mean value of elements of vector x.

```
min(x)
```
 minimum element of vector x.

```
pow2(x)
```
 2^x.

```
prod(x)
```
 product of the elements of x.

```
rand
```
 pseudo-random number in the interval (0, 1). The value returned is *pseudo*-random rather than truly random in the sense that there is an algorithm that determines rand from the initial 'seed'. The same seed will generate the same 'random' sequence; see a later chapter for how to seed rand by looking in the index.

```
realmax
```
 largest positive floating point number on your computer.

```
realmin
```
 smallest positive floating point number on your computer.

```
rem(x, y)
```
 remainder when x is divided by y, e.g., rem(19, 5) returns 4 (5 goes 3 times into 19, remainder 4).

 Strictly, rem(x, y) returns x - y * n where n = fix(x/y) is the integer nearest to x/y. This shows how negative and/or non-integer arguments are handled.

 fix and rem are useful for converting smaller units to larger ones, e.g., inches to feet and inches (one foot = 12 inches). The following statements convert 40 inches this way:

```
feet = fix(40/12)
inches = rem(40, 12)
```

Let us next examine an example that will hopefully inspire you to examine all of the functions listed as well as any other function that you may discover that MATLAB supports. Let us consider the arc-cosine, the arc-sine and the arc-tangent functions, i.e., acos(x), asin(x) and atan(x), respectively. If you specify x, i.e., the cosine, the sine and the tangent, respectively, between −1 and 1, what quadrant of the circle are the output angles selected? To provide an answer, the following *M-file* script was created and executed. The graphical comparison of the computed results is illustrated in Fig. 4.2. The REMARKS at the end of the script provides an interpretation of the graphical results and, hence, an answer to the question raised.

```
%
% Script to compare the acos(x), asin(x), and atan(x)
% functions over the range -1 < x < 1. The values are
% converted to angles in degrees. The results are
% compared graphically.
%
% Script prepared by D. T. Valentine - September 2006.
% Comments modified by D.T.V. ... 2008/2012/2016/2018.
%
% The question raised is: What range of angles, i.e.,
% which of the four quadrents of the circle from 0 to
% 2*pi are the angular outputs of each of the functions?
%
% Assign the values of x to be examined:
%
x = -1:0.001:1;
%
% Compute the arc-functions:
%
y1 = acos(x);
y2 = asin(x);
y3 = atan(x);
%
% Convert the angles from radians to degrees:
%
y1 = 180*y1/pi;
y2 = 180*y2/pi;
y3 = 180*y3/pi;
%
% Plot the results:
%
plot(y1,x,y2,x,y3,x),grid
legend('asin(x)', 'acos(x)', 'atan(x)')
xlabel('\theta in degrees')
ylabel('x, the argument of the function')
%
% REMARKS: Note the following:
%     (1) The acos(x) varies from 0 to 90 to 180 degrees.
%     (2) The asin(x) varies from -90 to 0 to 90 degrees.
%     (3) The atan(x) varies from -90 to 0 to 90 degrees.
%         To check remark (3) try atan(10000000) *180/pi.
%
% Stop
```

FIGURE 4.2
Comparison of the results from the acos, asin and atan functions.

4.2 IMPORTING AND EXPORTING DATA

When you get into serious programming you will often need to store data on a disk. The process of moving data between MATLAB and disk files is called importing (from disk files) and exporting (to disk files). Data are saved in disk files in one of two formats: text or binary. In text format, data values are ASCII codes, and can be viewed in any text editor. In binary format, data values are not ASCII codes and cannot be viewed in a text editor. Binary format is more efficient in terms of storage space required. This chapter provides a brief summary of the main ways in which MATLAB imports and exports data. For full details consult **MATLAB Help: Development Environment: Importing and Exporting Data**.

4.2.1 The load and save commands

If you want to save data between MATLAB sessions the save and load commands are probably the best ones to use.

4.2.2 Exporting text (ASCII) data

To export (save) the array

```
A =
    1    2    3
    4    5    6
```

in 'delimited' ASCII format in the file myData.txt use the command

```
save myData.txt A -ascii
```

If you view `myData.txt` in a text editor (or `type` it in the Command Window) it looks like this:

```
1.0000000e+000  2.0000000e+000  3.0000000e+000
4.0000000e+000  5.0000000e+000  6.0000000e+000
```

Delimiters are the characters used to separate the data values in the file—spaces by default. You can use tabs instead of spaces by specifying the `-tabs` qualifier instead of `-ascii`. If you save character arrays (strings) in this way, the ASCII codes of the characters are written to the file.

4.2.3 Importing text (ASCII) data

The `load` command is the reverse of `save`, but has a curious twist. If the array A has been saved in `myData.txt` as above the command

```
load myData.txt
```

creates a variable in the workspace *with the same name as the file, minus the extension*, i.e., `myData`. If you don't want the filename as the variable name use the functional form of the command, e.g.,

```
A = load('myData.txt')
```

Data imported in this way doesn't have to be created by MATLAB. You can create it in a text editor, or it could be created by any other program that exports data in ASCII format.

4.2.4 Exporting binary data

The command

```
save filename x y z
```

saves the variables `x`, `y` and `z` in the file *filename*.mat in MATLAB proprietary binary format, i.e., such a MAT-file can only be used by MATLAB.

Note:

- If no variables are listed the entire workspace is saved.
- The extension .mat is the default—you can specify a different extension.
- Seek `help` for all the `save` options.

4.2.5 Importing binary data

The command

> load *filename*

loads all the variables from *filename*.mat into the workspace; see help for all the load options.

SUMMARY

- MATLAB functions may be used to perform a variety of mathematical, trigonometric and other operations.
- Data can be saved to disk files in text (ASCII) format or in binary format.
- load and save can be used to import/export both text and binary data (the latter in the form of MAT-files).

EXERCISES

4.1 Write some MATLAB statements which will:
 (a) find the length C of the hypotenuse of a right-angle triangle in terms of the lengths A and B of the other two sides;
 (b) find the length C of a side of a triangle given the lengths A and B of the other two sides and the size in degrees of the included angle θ, using the cosine rule:

$$C^2 = A^2 + B^2 - 2AB\cos(\theta).$$

4.2 Translate the following formulae into MATLAB expressions:
 (a) $\ln(x + x^2 + a^2)$
 (b) $[e^{3t} + t^2\sin(4t)]\cos^2(3t)$
 (c) $4\tan^{-1}(1)$ (inverse tangent)
 (d) $\sec^2(x) + \cot(y)$
 (e) $\cot^{-1}(|x/a|)$ (use MATLAB's inverse cotangent)

4.3 There are 39.37 inches in a meter, 12 inches in a foot, and three feet in a yard. Write a script to input a length in meters (which may have a decimal part) and convert it to yards, feet and inches. (Check: 3.51 meters converts to 3 yds 2 ft 6.19 in.)

4.4 A sphere of mass m_1 impinges obliquely on a stationary sphere of mass m_2, the direction of the blow making an angle α with the line of motion of the impinging sphere. If the coefficient of restitution is e it can be shown that the impinging sphere is deflected through an angle β such that

$$\tan(\beta) = \frac{m_2(1 + e)\tan(\alpha)}{m_1 - em_2 + (m_1 + m_2)\tan^2(\alpha)}$$

Write a script to input values of m_1, m_2, e, and α (in degrees) and to compute the angle β in degrees.

4.5 Section 2.7.3 has a program for computing the members of the sequence $x_n = a^n/n!$. The program displays every member x_n computed. Adjust it to display only every 10th value of x_n.

Hint: the expression `rem(n, 10)` will be zero only when `n` is an exact multiple of 10. Use this in an `if` statement to display every tenth value of x_n.

4.6 To convert the variable `mins` minutes into hours and minutes you would use `fix(mins/60)` to find the whole number of hours, and `rem(mins, 60)` to find the number of minutes left over. Write a script which inputs a number of minutes and converts it to hours and minutes.

Now write a script to convert seconds into hours, minutes and seconds. Try out your script on 10 000 seconds, which should convert to 2 hours, 46 minutes and 40 seconds.

4.7 Design an algorithm (i.e., write the structure plan) for a machine which must give the correct amount of change from a $100 note for any purchase costing less than $100. The plan must specify the number and type of all notes and coins in the change, and should in all cases give as few notes and coins as possible. (If you are not familiar with dollars and cents, use your own monetary system.)

4.8 A uniform beam is freely hinged at its ends $x = 0$ and $x = L$, so that the ends are at the same level. It carries a uniformly distributed load of W per unit length, and there is a tension T along the x-axis. The deflection y of the beam a distance x from one end is given by

$$y = \frac{WEI}{T^2} \left[\frac{\cosh[a(L/2 - x)]}{\cosh(aL/2)} - 1 \right] + \frac{Wx(L - x)}{2T},$$

where $a^2 = T/EI$, E being the Young's modulus of the beam, and I is the moment of inertia of a cross-section of the beam. The beam is 10 m long, the tension is 1 000 N, the load 100 N/m, and EI is 10^4.

Write a script to compute and plot a graph of the deflection y against x (MATLAB has a cosh function).

To make the graph look realistic you will have to override MATLAB's automatic axis scaling with the statement

```
axis( [xmin xmax ymin ymax] )
```

after the `plot` statement, where `xmin` etc. have appropriate values.

Logical Vectors

THE OBJECTIVES OF THIS CHAPTER ARE TO ENABLE YOU TO:

- Understand logical operators more fully

And to introduce you to:

- Logical vectors and how to use them effectively in a number of applications
- Logical functions

CONTENTS

Examples 110
Discontinuous graphs 110
Avoiding division by zero 111
Avoiding infinity 112
Counting random numbers 113
Rolling dice 114

Logical operators 115
Operator precedence 116
Danger 116
Logical operators and vectors 117

Subscripting with logical vectors 118

Logical functions 119
Using any *and* all .. 120

Logical vectors instead of elseif ladders 121

Summary 123

Exercises 124

Supplementary material 126

This chapter introduces a most powerful and elegant feature of MATLAB, viz., the *logical vectors*. The topic is so useful and, hence, important that it deserves a chapter of its own.

Try these exercises on the command line:

1. Enter the following statements:

   ```
   r = 1;
   r <= 0.5     % no semi-colon
   ```

 If you correctly left out the semi-colon after the second statement you will have noticed that it returned the value 0.

2. Now enter the expression r >= 0.5 (again, no semi-colon). It should return the value 1. Note that we already saw in Chapter 2 that a logical expression in MATLAB involving only scalars returns a value of 0 if it is FALSE, and 1 if it is TRUE.

3. Enter the following statements:

   ```
   r = 1:5;
   r <= 3
   ```

 Now the logical expression r <= 3 (where r is a vector) returns a *vector*:

   ```
   1    1    1    0    0
   ```

 Can you see how to interpret this result? For each element of r for which r <= 3 is true, 1 is returned; otherwise 0 is returned. Now enter r == 4. Can you see why 0 0 0 1 0 is returned?

109

Essential MATLAB for Engineers and Scientists. https://doi.org/10.1016/B978-0-08-102997-8.00011-7

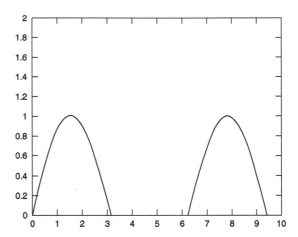

FIGURE 5.1
A discontinuous graph using logical vectors.

When a vector is involved in a logical expression, the comparison is carried out *element by element* (as in an arithmetic operation). If the comparison is true for a particular element of the vector, the resulting vector, which is called a *logical vector*, has a 1 in the corresponding position; otherwise it has a 0. The same applies to logical expressions involving matrices.

You can also compare vectors with vectors in logical expressions. Enter the following statements:

```
a = 1:5;
b = [0 2 3 5 6];
a == b       % no semi-colon!
```

The logical expression a == b returns the logical vector

```
0     1     1     0     0
```

because it is evaluated element by element, i.e., a(1) is compared with b(1), a(2) with b(2), etc.

5.1 EXAMPLES

5.1.1 Discontinuous graphs

One very useful application of logical vectors is in plotting discontinuities. The following script plots the graph, shown in Fig. 5.1, defined by

$$y(x) = \begin{cases} \sin(x) & (\sin(x) > 0) \\ 0 & (\sin(x) \le 0) \end{cases}$$

over the range 0 to 3π:

```
x = 0 : pi/20 : 3 * pi;
y = sin(x);
y = y .* (y > 0);   % set negative values of sin(x) to zero
plot(x, y)
```

The expression y > 0 returns a logical vector with 1's where sin(x) is positive, and 0's otherwise. Element-by-element multiplication by y with .* then picks out the positive elements of y.

5.1.2 Avoiding division by zero

Suppose you want to plot the graph of $\sin(x)/x$ over the range -4π to 4π. The most convenient way to set up a vector of the x co-ordinates is

```
x = -4*pi : pi / 20 : 4*pi;
```

But then when you try

```
y = sin(x) ./ x;
```

you get for the value of y at $x = 0$ NaN. A neat way around this problem is to use a logical vector to replace the zero with eps. This MATLAB function returns the difference between 1.0 and the next largest number which can be represented in MATLAB, i.e., approximately 2.2e-16. Here is how to do it:

```
x = x + (x == 0)*eps;
```

The expression x == 0 returns a logical vector with a single 1 for the element of x which is zero, and so eps is added only to that element. The following script plots the graph correctly—without a missing segment at $x = 0$ (see Fig. 5.2).

```
x = -4*pi : pi/20 : 4*pi;
x = x + (x == 0)*eps;        % adjust x = 0 to x = eps
y = sin(x) ./ x;
plot(x, y)
```

When x has the value eps, the value of sin(eps)/eps has the correct limiting value of 1 (check it), instead of NaN (Not-a-Number) resulting from a division by zero.

FIGURE 5.2
$\sin(x)/x$.

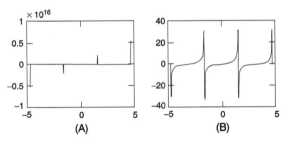

FIGURE 5.3
Variations on $\tan(x)$: (A) Unrestricted vertical coordinate. (B) Restricted vertical coordinate.

5.1.3 Avoiding infinity

The following script attempts to plot $\tan(x)$ over the range $-3\pi/2$ to $3\pi/2$. If you are not too hot at trig graphs, perhaps you should sketch the graph roughly with pen and paper before you run the script!

```
x = -3/2*pi : pi/100 : 3/2*pi;
y = tan(x);
plot(x, y)
```

The MATLAB plot—the left graph in Fig. 5.3—should look nothing like your sketch. The problem is that $\tan(x)$ approaches $\pm\infty$ at odd multiples of $\pi/2$. The scale on the MATLAB plot is therefore very large (about 10^{15}), making it impossible to see the structure of the graph anywhere else.

If you add the statement

```
y = y .* (abs(y) < 1e10);   % remove the big ones
```

just before the `plot` statement you'll get a much nicer graph, as shown in Fig. 5.3B. The expression `abs(y) < 1e10` returns a logical vector which is zero only at the asymptotes. The graph thus goes through zero at these points, which incidentally draws nearly vertical asymptotes for you. These 'asymptotes' become more vertical as the increment in x becomes smaller.

5.1.4 Counting random numbers

The function `rand` returns a (pseudo-)random number in the interval $[0, 1)$; `rand(1, n)` returns a row vector of n such numbers. Work out the following problem on the command line:

1. Set up a vector r with seven random elements (leave out the semi-colon so that you can see its elements):

   ```
   r = rand(1,7)      % no semi-colon
   ```

 Check that the logical expression `r < 0.5` gives the correct logical vector.
2. Using the function `sum` on the logical expression `r < 0.5` will effectively count how many elements of r are less than 0.5. Try it, and check your answer against the values displayed for r:

   ```
   sum( r < 0.5 )
   ```

3. Now use a similar statement to count how many elements of r are greater than or equal to 0.5 (the two answers should add up to 7, shouldn't they?).
4. Since `rand` generates *uniformly distributed* random numbers, you would expect the number of random numbers which are less than 0.5 to get closer and closer to half the total number as more and more are generated.

 Generate a vector of a few thousand random numbers (suppress display with a semi-colon this time) and use a logical vector to count how many are less than 0.5.

 Repeat a few times with a new set of random numbers each time. Because the numbers are random, you should never get quite the same answer each time.

Without logical vectors this problem is a little more involved. Here is the program:

```
tic                      % start
a = 0;                   % number >= 0.5
b = 0;                   % number < 0.5

for n = 1:5000
  r = rand;              % generate one number per loop
  if r >= 0.5
    a = a + 1;
  else
    b = b + 1;
  end;
end;

t = toc;                 % finish
disp( ['less than 0.5: ' num2str(a)] )
disp( ['time: ' num2str(t)] )
```

It also takes a lot longer. Compare times for the two methods on your computer.

5.1.5 Rolling dice

When a fair dice is rolled, the number uppermost is equally likely to be any integer from 1 to 6. So if rand is a random number in the range [0, 1), 6 * rand will be in the range [0, 6), and 6 * rand + 1 will be in the range [1, 7), i.e., between 1 and 6.9999. Discarding the decimal part of this expression with floor gives an integer in the required range. Try the following exercises:

1. Generate a vector d of 20 random integers in the range 1 to 6:

    ```
    d = floor(6 * rand(1, 20)) + 1
    ```

2. Count the number of 'sixes' thrown by summing the elements of the logical vector d == 6.
3. Verify your result by displaying d.
4. Estimate the probability of throwing a six by dividing the number of sixes thrown by 20. Using random numbers like this to mimic a real situation based on chance is called *simulation*.
5. Repeat with more random numbers in the vector d. The more you have, the closer the proportion of sixes gets to the theoretical expected value of 0.1667, i.e., 1/6.
6. Can you see why it would be incorrect to use round instead of floor? The problem is that round rounds in both directions, whereas floor rounds everything down.

Table 5.1 Logical operators.

Operator	Meaning
~	NOT
&	AND
\|	OR

Table 5.2 Truth table (T = true; F = false).

lex1	lex2	~lex1	lex1 & lex2	lex1 \| lex2	xor(lex1, lex2)
F	F	T	F	F	F
F	T	T	F	T	T
T	F	F	F	T	T
T	T	F	T	T	F

5.2 LOGICAL OPERATORS

We saw briefly in Chapter 2 that logical expressions can be constructed not only from the six relational operators, but also from the three *logical operators* shown in Table 5.1. Table 5.2 shows the effects of these operators on the general logical expressions *lex1* and *lex2*.

The OR operator (|) is technically an *inclusive* OR, because it is true when either or both of its operands are true. MATLAB also has an *exclusive* OR function xor(a, b) which is 1 (true) only when either *but not both of* a and b is 1 (Table 5.2).

MATLAB also has a number of functions that perform bitwise logical operations. See Help on ops.

The precedence levels of the logical operators, among others, are shown in Table 5.3. As usual, precedences may be overridden with brackets, e.g.,

```
~ 0 & 0
```

returns 0 (false), whereas

```
~ (0 & 0)
```

returns 1 (true). Some more examples:

```
(b * (b == 4) * a * c) & (a ~= 0)
(final >= 60) & (final < 70)
(a ~= 0) | (b ~= 0) | (c != 0)
~((a == 0) & (b == 0) & (c == 0))
```

Table 5.3 Operator precedence (see **Help** on operator precedence).

Precedence	Operators	
1.	()	
2.	^ .^ ' .' (pure transpose)	
3.	+ (unary plus) - (unary minus) ~ (NOT)	
4.	* / \ .* ./ .\	
5.	+ (addition) - (subtraction)	
6.	:	
7.	> < >= <= == ~=	
8.	& (AND)	
9.		(OR)

It is never wrong to use brackets to make the logic clearer, even if they are syntactically unnecessary. Incidentally, the last two expressions above are logically equivalent, and are false only when a = b = c = 0. It makes you think, doesn't it?

5.2.1 Operator precedence

You may accidentally enter an expression like

```
2 > 1 & 0
```

(try it) and be surprised because MATLAB (a) accepts it, and (b) returns a value of 0 (false). It is surprising because:

(a) `2 > 1 & 0` doesn't appear to make sense. If you have got this far you deserve to be let into a secret. MATLAB is based on the notorious language C, which allows you to mix different types of operators in this way (Pascal, for example, would never allow such flexibility!).

(b) We instinctively feel that `&` should have the higher precedence. `1 & 0` evaluates to 0, so `2 > 0` should evaluate to 1 instead of 0. The explanation is due partly to the resolution of surprise (a). MATLAB groups its operators in a rather curious and non-intuitive way. The complete operator precedence is given in Table 5.3 (reproduced for ease of reference in Appendix A). (Recall that the transpose operator ' performs a *complex conjugate* transpose on complex data; the dot-transpose operator .' performs a 'pure' transpose without taking the complex conjugate.) Brackets always have the highest precedence.

5.2.2 Danger

I have seen quite a few students incorrectly convert the mathematical inequality $0 < r < 1$, say, into the MATLAB expression

```
0 < r < 1
```

Once again, the first time I saw this I was surprised that MATLAB did not report an error. Again, the answer is that MATLAB doesn't really mind how you mix up operators in an expression. It simply churns through the expression according to its rules (which may not be what you expect).

Suppose r has the value 0.5. Mathematically, the inequality is true for this value of r since it lies in the required range. However, the expression $0 < r < 1$ is evaluated as 0. This is because the left hand operation ($0 < 0.5$) is first evaluated to 1 (true), followed by $1 < 1$ which is false.

Inequalities like this should rather be coded as

```
(0 < r) & (r < 1)
```

The brackets are not strictly necessary (see Table 5.3) but they certainly help to clarify the logic.

5.2.3 Logical operators and vectors

The logical operators can also operate on vectors (of the same size), returning logical vectors, e.g.,

```
~(~[1 2 0 -4 0])
```

replaces all the non-zeros by 1's, and leaves the 0's untouched. Try it.

The script in Section 5.1 that avoids division by zero has the critical statement

```
x = x + (x == 0)*eps;
```

This is equivalent to

```
x = x + (~x)*eps;
```

Try it, and make sure you understand how it works.

Exercises

Work out the results of the following expressions before checking them at the command line:

```
a = [-1 0 3];
b = [0 3 1];
~a
a & b
```

```
a | b
xor(a, b)
a > 0 & b > 0
a > 0 | b > 0
~ a > 0
a + (~ b)
a > ~ b
~ a > b
~ (a > b)
```

5.3 SUBSCRIPTING WITH LOGICAL VECTORS

We saw briefly in Chapter 2 that elements of a vector may be referenced with subscripts, and that the subscripts themselves may be vectors, e.g.,

```
a = [-2 0 1 5 9];
a([5 1 3])
```

returns

```
9    -2    1
```

i.e., the fifth, first and third elements of a. In general, if x and v are vectors, where v has *n* elements, then x(v) means

```
[x(v(1)), x(v(2)), ..., x(v(n))]
```

Now with a as defined above, see what

```
a(logical([0 1 0 1 0]))
```

returns. The function logical(v) returns a logical vector, with elements which are 1 or 0 according as the elements of v are non-zero or 0.

A summary of the rules for the use of a logical vector as a subscript are as follows:

- A logical vector v may be a subscript of another vector x.
- Only the elements of x corresponding to 1's in v are returned.
- x and v must be the same size.

Thus, the statement above returns

```
0    5
```

i.e., the second and fourth elements of a, corresponding to the 1's in logical([0 1 0 1 0]).

What will

```
a(logical([1 1 1 0 0]))
```

return? And what about a(logical([0 0 0 0 0]))?

Logical vector subscripts provide an elegant way of removing certain elements from a vector, e.g.,

```
a = a(a > 0)
```

removes all the non-positive elements from a, because a > 0 returns the logical vector [0 0 1 1 1]. We can verify incidentally that the expression a > 0 is a logical vector, because the statement

```
islogical(a > 0)
```

returns 1. However, the *numeric* vector [0 0 1 1 1] is not a logical vector; the statement

```
islogical([0 0 1 1 1])
```

returns 0.

5.4 LOGICAL FUNCTIONS

MATLAB has a number of useful *logical functions* that operate on scalars, vectors and matrices. Examples are given in the following list (where x is a vector unless otherwise stated). See Help on logical functions. (The functions are defined slightly differently for matrix arguments—see Chapter 6 or Help.)

any(x) returns the scalar 1 (true) if *any* element of x is non-zero (true).

all(x) returns the scalar 1 if *all* the elements of x are non-zero.

exist('a') returns 1 if a is a workspace variable. For other possible return values see help. Note that a must be enclosed in apostrophes.

find(x) returns a vector containing the subscripts of the *non-zero* (true) elements of x, so for example,

```
a = a( find(a) )
```

removes all the zero elements from a! Try it. Another use of find is in finding the subscripts of the largest (or smallest) elements in a vector, when there is more than one. Enter the following:

```
x = [8 1 -4 8 6];
find(x >= max(x))
```

This returns the vector [1 4], which are the subscripts of the largest element (8). It works because the logical expression x >= max(x) returns a logical vector with 1's only at the positions of the largest elements.

isempty(x) returns 1 if x is an empty array and 0 otherwise. An empty array has a size of 0-by-0.

isinf(x) returns 1's for the elements of x which are +Inf or −Inf, and 0's otherwise.

isnan(x) returns 1's where the elements of x are NaN and 0's otherwise. This function may be used to remove NaNs from a set of data. This situation could arise while you are collecting statistics; missing or unavailable values can be temporarily represented by NaNs. However, if you do any calculations involving the NaNs, they propagate through intermediate calculations to the final result. To avoid this, the NaNs in a vector may be removed with a statement like

```
x(isnan(x)) = [ ]
```

MATLAB has a number of other logical functions starting with the characters is. See is* in the Help index for the complete list.

5.4.1 Using any **and** all

Because any and all with vector arguments return *scalars*, they are particularly useful in if statements. For example,

```
if all(a >= 1)
    do something
end
```

means 'if all the elements of the vector a are greater than or equal to 1, then *do something*'.

Recall from Chapter 2 that a vector condition in an if statement is true only if *all* its elements are non-zero. So if you want to execute *statement* below when two vectors a and b are equal (i.e., the same) you could say

```
if a == b
    statement
end
```

since if considers the logical vector returned by a == b true only if every element is a 1.

If, on the other hand, you want to execute *statement* specifically when the vectors a and b are *not* equal, the temptation is to say

```
    if a ~= b      % wrong wrong wrong!!!
    statement
end
```

However this will *not* work, since *statement* will only execute if *each* of the corresponding elements of a and b differ. This is where any comes in:

```
    if any(a ~= b)     % right right right!!!
    statement
end
```

This does what is required since any(a ~= b) returns the scalar 1 if *any* element of a differs from the corresponding element of b.

5.5 LOGICAL VECTORS INSTEAD OF elseif LADDERS

Those of us who grew up on more conventional programming languages in the last century may find it difficult to think in terms of using logical vectors when solving general problems. It's a nice challenge whenever writing a program to ask yourself whether you can possibly use logical vectors. They are almost always faster than other methods, although often not as clear to read later. You must decide when it's important for you to use logical vectors. But it's a very good programming exercise to force yourself to use them whenever possible! The following example illustrates these points by first solving a problem conventionally, and then with logical vectors.

It has been said that there are two unpleasant and unavoidable facts of life: death and income tax. A very simplified version of how income tax is calculated could be based on the following table:

Taxable income	Tax payable
$10 000 or less	10 per cent of taxable income
between $10 000 and $20 000	$1000 + 20 per cent of amount by which taxable income exceeds $10 000
more than $20 000	$3000 + 50 per cent of amount by which taxable income exceeds $20 000

The tax payable on a taxable income of $30 000, for example, is

$3000 + 50 per cent of ($30 000 − $20 000), i.e., $8000.

We would like to calculate the income tax on the following taxable incomes (in dollars): 5000, 10 000, 15 000, 30 000 and 50 000.

The conventional way to program this problem is to set up a vector with the taxable incomes as elements and to use a `loop` with an `elseif` ladder to process each element, as follows:

```
% Income tax the old-fashioned way

inc = [5000 10000 15000 30000 50000];

for ti = inc

  if ti < 10000
    tax = 0.1 * ti;
  elseif ti < 20000
    tax = 1000 + 0.2 * (ti - 10000);
  else
    tax = 3000 + 0.5 * (ti - 20000);
  end;

  disp( [ti tax] )
end;
```

Here is the output, suitably edited (note that the amount of tax paid changes continuously between tax 'brackets'—each category of tax is called a bracket):

Taxable income	Income tax
5000.00	500.00
10000.00	1000.00
15000.00	2000.00
30000.00	8000.00
50000.00	18000.00

Now here is the logical way:

```
% Income tax the logical way

inc = [5000 10000 15000 30000 50000];

tax = 0.1 * inc .* (inc <= 10000);
tax = tax + (inc > 10000 & inc <= 20000) ...
                        .* (0.2 * (inc-10000) + 1000);
tax = tax + (inc > 20000) .* (0.5 * (inc-20000) + 3000);

disp( [inc' tax'] );
```

To understand how it works, it may help to enter the statements on the command line. Start by entering the vector `inc` as given. Now enter

```
inc <= 10000
```

which should give the logical vector [1 1 0 0 0]. Next enter

```
inc .* (inc <= 10000)
```

which should give the vector [5000 10000 0 0 0]. This has successfully selected only the incomes in the first bracket. The tax for these incomes is then calculated with

```
tax = 0.1 * inc .* (inc <= 10000)
```

which returns [500 1000 0 0 0].

Now for the second tax bracket. Enter the expression

```
inc > 10000 & inc <= 20000
```

which returns the logical vector [0 0 1 0 0], since there is only one income in this bracket. Now enter

```
0.2 * (inc-10000) + 1000
```

This returns [0 1000 2000 5000 9000]. Only the third entry is correct. Multiplying this vector by the logical vector just obtained blots out the other entries, giving [0 0 2000 0 0]. The result can be safely added to the vector `tax` since it will not affect the first two entries already there.

I am sure you will agree that the logical vector solution is more interesting than the conventional one!

SUMMARY

- When a relational and/or logical operator operates on a vector expression, the operation is carried out element by element. The result is a logical vector consisting of 0's (FALSE) and 1's (TRUE).
- A vector may be subscripted with a logical vector of the same size. Only the elements corresponding to the 1's in the logical vector are returned.
- When one of the logical operators (~ & |) operates on an expression any non-zero value in an operand is regarded as TRUE; zero is regarded as FALSE. A logical vector is returned.

- Arithmetic, relational and logical operators may appear in the same expression. Great care must be taken in observing the correct operator precedence in such situations.
- Vectors in a logical expression must all be the same size.
- If a logical expression is a vector or a matrix, it is considered true in an `if` statement only if all its elements are non-zero.
- The logical functions `any` and `all` return scalars when taking vector arguments, and are consequently useful in `if` statements.
- Logical vectors may often be used instead of the more conventional `elseif` ladder. This provides faster more elegant code, but requires more ingenuity and the code may be less clear to read later on.

EXERCISES

5.1 Determine the values of the following expressions yourself before checking your answers using MATLAB. You may need to consult Table 5.3.

(a) `1 & -1`
(b) `13 & ~(-6)`
(c) `0 < -2|0`
(d) `~[1 0 2] * 3`
(e) `0 <= 0.2 <= 0.4`
(f) `5 > 4 > 3`
(g) `2 > 3 & 1`

5.2 Given that `a = [1 0 2]` and `b = [0 2 2]` determine the values of the following expressions. Check your answers with MATLAB.

(a) `a ~= b`
(b) `a < b`
(c) `a < b < a`
(d) `a < b < b`
(e) `a | (~a)`
(f) `b & (~b)`
(g) `a(~(~b))`
(h) `a = b == a` (determine final value of `a`)

5.3 Write some MATLAB statements on the command line which use logical vectors to count how many elements of a vector `x` are negative, zero or positive. Check that they work, e.g., with the vector

```
[-4  0  5  -3  0  3  7  -1  6]
```

5.4 The Receiver of Revenue (Internal Revenue Service) decides to change the tax table used in Section 5.5 slightly by introducing an extra tax bracket and changing the tax-rate in the third bracket, as shown in the table on the next page.

Taxable income	Tax payable
$10 000 or less	10 per cent of taxable income
between $10 000 and $20 000	$1000 + 20 per cent of amount by which taxable income exceeds $10 000
between $20 000 and $40 000	$3000 + 30 per cent of amount by which taxable income exceeds $20 000
more than $40 000	$9000 + 50 per cent of amount by which taxable income exceeds $40 000

Amend the logical vector script to handle this table, and test it on the following list of incomes (dollars): 5000, 10 000, 15 000, 22 000, 30 000, 38 000 and 50 000.

5.5 A certain company offers seven annual salary levels (dollars): 12 000, 15 000, 18 000, 24 000, 35 000, 50 000 and 70 000. The number of employees paid at each level are, respectively: 3000, 2500, 1500, 1000, 400, 100 and 25. Write some statements at the command line to find the following:

(a) The average salary *level*. Use `mean`. (Answer: 32 000)

(b) The number of employees above and below this average salary level. Use logical vectors to find which salary levels are above and below the average level. Multiply these logical vectors element by element with the employee vector, and `sum` the result. (Answer: 525 above, 8000 below)

(c) The *average salary earned* by an individual in the company (i.e., the total annual salary bill divided by the total number of employees). (Answer: 17 038.12).

5.6 Write some statements on the command line to remove the largest element(s) from a vector. Try it out on x = [1 2 5 0 5]. The idea is to end up with [1 2 0] in x. Use `find` and the empty vector [].

5.7 The electricity accounts of residents in a very small rural community are calculated as follows:

- if 500 units or less are used the cost is 2 cents per unit;
- if more than 500, but not more than 1000 units are used, the cost is $10 for the first 500 units, and then 5 cents for every unit in excess of 500;
- if more than 1000 units are used, the cost is $35 for the first 1000 units plus 10 cents for every unit in excess of 1000;
- in addition, a basic service fee of $5 is charged, no matter how much electricity is used.

The five residents use the following amounts (units) of electricity in a certain month: 200, 500, 700, 1000 and 1500. Write a program which

uses logical vectors to calculate how much they must pay. Display the results in two columns: one for the electricity used in each case, and one for amount owed. (Answers: $9, $15, $25, $40, $90)

APPENDIX 5.A SUPPLEMENTARY MATERIAL

Supplementary material related to this chapter can be found online at https://doi.org/10.1016/B978-0-08-102997-8.00011-7.

Matrices and Arrays

THE OBJECTIVES OF THIS CHAPTER ARE TO:

- Introduce you to ways of creating and manipulating matrices.
- Introduce you to matrix operations.
- Introduce you to character strings and facilities for handling them.
- Demonstrate some uses of matrices by examples in:
 - Population dynamics.
 - Markov processes.
 - Linear equations.
- Introduce MATLAB's sparse matrix facilities.

CONTENTS

Matrices.......... 127
A concrete example 127
Creating matrices .. 129
Subscripts 129
Transpose 130
The colon operator . 130
*Duplicating rows
and columns: Tiling* 133
*Deleting rows and
columns...............* 134
*Elementary
matrices* 135
*Specialized
matrices* 136
*Using MATLAB
functions with
matrices* 137
*Manipulating
matrices* 138
*Array (element-by-
element) operations
on matrices* 138
Matrices and for *...* 139
*Visualization of
matrices* 139
*Vectorizing
nested* fors*: loan
repayment tables...* 139
*Multi-dimensional
arrays* 142

Matrix
operations 143
*Matrix
multiplication........* 143
*Matrix
exponentiation.......* 145

Other matrix
functions 145 **127**

As we have already seen, the name MATLAB stands for MATrix LABoratory, because the MATLAB system is designed specially to work with data arranged in the form of *matrices* (2-D arrays). The word 'matrix' has two distinct meanings in this chapter:

1. An arrangement of data in rows and columns, e.g., a table.
2. A mathematical object, for which particular mathematical operations are defined, e.g., 'matrix' multiplication.

We begin this chapter (in Sections 6.1, 6.2, and 6.3) by looking at matrices in the first sense, summarizing and extending what we learned about them in Chapter 2. We then go on to look briefly at the mathematical operations on matrices. In a later chapter we will see how these operations can be applied in a number of widely differing areas, such as systems of linear equations, population dynamics, and Markov processes.

6.1 MATRICES

6.1.1 A concrete example

A ready-mix concrete company has three factories (S1, S2 and S3) which must supply three building sites (D1, D2 and D3). The costs, in some suitable cur-

Essential MATLAB for Engineers and Scientists. https://doi.org/10.1016/B978-0-08-102997-8.00012-9

Population growth: Leslie matrices 146

Markov processes 149
A random walk 150

Linear equations 152
MATLAB's solution . 153
The residual 153
Over-determined systems 154
Under-determined systems 155
Ill conditioning 155
Matrix division 156

Sparse matrices 157

Summary 160

Exercises 160

rency, of transporting a load of concrete from any factory to any site are given by the following cost table:

	D1	D2	D3
S1	3	12	10
S2	17	18	35
S3	7	10	24

The factories can supply 4, 12 and 8 loads per day, respectively, and the sites require 10, 9 and 5 loads per day, respectively. The real problem is to find the cheapest way to satisfy the demands at the sites, but we are not considering that here.

Suppose the factory manager proposes the following transportation scheme (each entry represents the number of loads of concrete to be transported along that particular route):

	D1	D2	D3
S1	4	0	0
S2	6	6	0
S3	0	3	5

This sort of scheme is called a *solution* to the transportation problem. The cost table (and the solution) can then be represented by tables **C** and **X**, say, where c_{ij} is the entry in row i and column j of the cost table, with a similar convention for **X**.

To compute the cost of the above solution, each entry in the solution table must be multiplied by the corresponding entry in the cost table. (This operation is not to be confused with the mathematical operation of matrix multiplication, which is discussed later.) We therefore want to calculate

$$3 \times 4 + 12 \times 0 + \cdots + 24 \times 5.$$

To do this calculation in MATLAB, enter **C** and **X** as matrices from the command line, with a semi-colon at the end of each row:

```
c = [3 12 10; 17 18 35; 7 10 24];
x = [4 0 0; 6 6 0; 0 3 5];
```

and then find the *array* product of c and x:

```
total = c .* x
```

which gives

```
     12     0     0
    102   108     0
      0    30   120
```

The command

```
sum(total)
```

then returns a vector where each element is the sum of each column of total:

```
114    138    120
```

Summing this in turn, i.e., sum(sum(total)), gives the final answer of 372.

6.1.2 Creating matrices

As we saw above, use a semi-colon to indicate the end of a row when entering a matrix. Bigger matrices can be constructed from smaller ones, e.g., the statements

```
a = [1 2; 3 4];
x = [5 6];
a = [a; x]
```

result in

```
a =
      1     2
      3     4
      5     6
```

Instead of a semi-colon, you can use a line-feed (**Enter**) to indicate the end of a row.

6.1.3 Subscripts

Individual elements of a matrix are referenced with two subscripts, the first for the row, and the second for the column, e.g., a(3,2) for a above returns 6.

Alternatively, and less commonly, a single subscript may be used. In this case you can think of the matrix as being 'unwound' column by column. So a(5) for the above example returns 4.

If you refer to a subscript which is out of range, e.g., a(3,3) for a above, you will get an error message. However, if you assign a value to an element with a subscript which is out of range the matrix is enlarged to accommodate the new element, for example, the assignment

```
a(3,3) = 7
```

will add a third column to `a` with 0's everywhere except at `a(3,3)`.

6.1.4 Transpose

The statements

```
a = [1:3; 4:6]
b = a'
```

result in

```
a =
        1       2       3
        4       5       6

b =
        1       4
        2       5
        3       6
```

The transpose operator ' (apostrophe) turns rows into columns and vice versa.

6.1.5 The colon operator

- The colon operator is extremely powerful, and provides for very efficient ways of handling matrices, e.g., if `a` is the matrix

```
a =
        1       2       3
        4       5       6
        7       8       9
```

the statement

```
a(2:3,1:2)
```

results in

```
        4       5
        7       8
```

(returns second and third rows, first and second columns), the statement

```
a(3,:)
```

results in

```
     7     8     9
```

(returns third row), and the statement

```
a(1:2,2:3) = ones(2)
```

results in

```
a =
     1     1     1
     4     1     1
     7     8     9
```

(replaces the 2-by-2 submatrix composed of the first and second row and the second and third column with a square matrix of 1's).

Essentially, what is happening in the above examples is that the colon operator is being used to create vector subscripts. However, a colon by itself in place of a subscript denotes *all* the elements of the corresponding row or column. So a(3,:) means all elements in the third row.

This feature may be used, for example, to construct tables. Suppose we want a table trig of the sines and cosines of the angles 0° to 180° in steps of 30°. The following statements achieve this:

```
x = [0:30:180]';
trig(:,1) = x;
trig(:,2) = sin(pi/180*x);
trig(:,3) = cos(pi/180*x);
```

■ You can use vector subscripts to get more complicated effects, e.g., for

```
b = ones(3,4)
```

try

```
a(:,[1 3]) = b(:,[4 2])
```

This operation replaces the first and third columns of a by the fourth and second columns of b (a and b must have the same number of rows).

■ The colon operator is ideal for the sort of row operations performed in Gauss reduction (a technique of numerical mathematics), for example, if a is the matrix

```
a =
     1    -1     2
     2     1    -1
     3     0     1
```

the statement

```
a(2,:) = a(2,:) - a(2,1)*a(1,:)
```

subtracts the first row multiplied by the first element in the second row from the second row, resulting in

```
a =
      1    -1     2
      0     3    -5
      3     0     1
```

(the idea being to get a zero immediately underneath $a(1,1)$).

- The keyword end refers to the *last* row or column of an array, for example, if r is a row vector, the statement

```
sum(r(3:end))
```

returns the sum of all the elements of r from the third one to the last one.

- The colon operator may also be used as a single subscript, in which case it behaves differently if it is on the right-hand or left-hand side of an assignment. On the right-hand side of an assignment, a(:) gives all the elements of a strung out *by columns* in one long column vector, e.g., if

```
a =
      1     2
      3     4
```

the statement

```
b = a(:)
```

results in

```
b =
      1
      3
      2
      4
```

However, on the left-hand side of an assignment, a(:) *reshapes* a matrix. a must already exist. Then a(:) denotes a matrix with the same dimensions (shape) as a, but with new contents taken from the right-hand side. In other words, the matrix on the right-hand side is reshaped into the shape of a on the left-hand side. Some examples may help. If

```
b =
      1     2     3
      4     5     6
```

and

```
a =
      0     0
      0     0
      0     0
```

the statement

```
a(:) = b
```

results in

```
a =
     1    5
     4    3
     2    6
```

(the contents of b are strung out into one long column and then fed into a by columns). As another example, the statement

```
a(:) = 1:6
```

(with a as above) results in

```
a =
     1    4
     2    5
     3    6
```

Reshaping can also be done with the reshape function. See help.
Whether the colon operator appears on the right- or left-hand side of an assignment, the matrices or submatrices on each side of the assignment have the same shape (except in the special case below).

■ As a special case, a single colon subscript may be used to replace *all* the elements of a matrix with a scalar, e.g.,

```
a(:) = -1
```

6.1.6 Duplicating rows and columns: Tiling

Sometimes it is useful to generate a matrix where all the rows or columns are the same. This can be done with the repmat function as follows. If a is the row vector

```
a =
     1    2    3
```

the statement

```
repmat(a, [3 1])
```

results in

```
ans =
     1     2     3
     1     2     3
     1     2     3
```

In `help`'s inimitable phraseology, this statement produces a 3-by-1 'tiling' of copies of `a`. You can think of `a` as a 'strip' of three tiles stuck to self-adhesive backing. The above statement tiles a floor with three rows and one column of such a strip of tiles.

There is an alternative syntax for `repmat`:

```
repmat(a, 3, 1)
```

An interesting example of this process appears at the end of this section in the loan repayment problem.

6.1.7 Deleting rows and columns

Use the colon operator and the empty array to delete entire rows or columns, for example,

```
a(:,2) = [ ]
```

deletes the second column of `a`.

You can't delete a single element from a matrix while keeping it a matrix, so a statement like

```
a(1,2) = [ ]
```

results in an error. However, using the single subscript notation you can delete a sequence of elements from a matrix and reshape the remaining elements into a row vector, for example, if

```
a =
     1     2     3
     4     5     6
     7     8     9
```

the statement

```
a(2:2:6) = [ ]
```

results in

```
a =
    1    7    5    3    6    9
```

Get it? (First unwind `a` by columns, then remove elements 2, 4 and 6.)

You can use logical vectors to extract a selection of rows or columns from a matrix, for example, if `a` is the original 3-by-3 matrix defined above, the statement

```
a(:, logical([1 0 1]))
```

results in

```
ans =
    1    3
    4    6
    7    9
```

(first and third columns extracted). The same effect is achieved with

```
a(:, [1 3])
```

6.1.8 Elementary matrices

There is a group of functions to generate 'elementary' matrices which are used in a number of applications. See `help elmat`.

For example, the functions `zeros`, `ones` and `rand` generate matrices of 1's, 0's and random numbers, respectively. With a single argument n, they generate $n \times n$ (square) matrices. With two arguments n and m they generate $n \times m$ matrices. (For very large matrices `repmat` is usually faster than `ones` and `zeros`.)

The function `eye(n)` generates an $n \times n$ *identity* matrix, i.e., a matrix with 1's on the main 'diagonal', and 0's everywhere else, e.g., the statement

```
eye(3)
```

results in

```
ans =
    1    0    0
    0    1    0
    0    0    1
```

(The original version of MATLAB could not use the more obvious name `I` for the identity since it did not distinguish between upper- and lowercase letters and `i` was the natural choice for the imaginary unit number.)

As an example, eye may be used to construct a *tridiagonal* matrix as follows. The statements

```
a = 2 * eye(5);
a(1:4, 2:5) = a(1:4, 2:5) - eye(4);
a(2:5, 1:4) = a(2:5, 1:4) - eye(4)
```

result in

```
a =
     2    -1     0     0     0
    -1     2    -1     0     0
     0    -1     2    -1     0
     0     0    -1     2    -1
     0     0     0    -1     2
```

Incidentally, if you work with large tridiagonal matrices, you should have a look at the sparse matrix facilities available in MATLAB via the help browser.

6.1.9 Specialized matrices

The following are functions that could be used to generate arbitrary matrices to investigate matrix operations when you cannot think of one to generate yourself. They are special matrices that were discovered by mathematicians, the motivation for their discovery and other details about them you are *not* expected to understand.

pascal(n) generates a Pascal matrix of order *n*. Technically, this is a symmetric positive definite matrix with entries made up from *Pascal's triangle*, e.g.,

```
pascal(4)
```

results in

```
ans =
     1     1     1     1
     1     2     3     4
     1     3     6    10
     1     4    10    20
```

magic(n) generates an *n* × *n* *magic square*.

MATLAB has a number of other functions which generate special matrices, such as gallery, hadamard, hankel, hilb, toeplitz, vander, etc. See help elmat.

6.1.10 Using MATLAB functions with matrices

When a MATLAB mathematical or trigonometric function has a matrix argument, the function operates on every element of the matrix, as you would expect. However, many of the other MATLAB functions operate on matrices *column by column*, e.g., if

```
a =
      1      0      1
      1      1      1
      0      0      1
```

the statement

```
all(a)
```

results in

```
ans =
      0      0      1
```

For each *column* of a where all the elements are true (non-zero) `all` returns 1, otherwise it returns 0. `all` therefore returns a logical vector when it takes a matrix argument. To test if all the elements of a are true, use `all` twice. In this example, the statement

```
all(all(a))
```

returns 0 because some of the elements of a are 0. On the other hand the statement

```
any(a)
```

returns

```
ans =
      1      1      1
```

because each column of a has at least one non-zero element, and `any(any(a))` returns 1, since a itself has at least one non-zero element.

If you are not sure whether a particular function operates columnwise or element by element on matrices, you can always request `help`.

6.1.11 Manipulating matrices

Here are some functions for manipulating matrices. See `help` for details.

`diag` extracts or creates a diagonal.

`fliplr` flips from left to right.

`flipud` flips from top to bottom.

`rot90` rotates.

`tril` extracts the lower triangular part, e.g., the statement

```
tril(pascal(4))
```

results in

```
ans =
     1     0     0     0
     1     2     0     0
     1     3     6     0
     1     4    10    20
```

`triu` extracts the upper triangular part.

6.1.12 Array (element-by-element) operations on matrices

The array operations discussed in Chapter 2 all apply to matrices as well as vectors. For example, if a is a matrix, a * 2 multiplies each element of a by 2, and if

```
a =
     1     2     3
     4     5     6
```

the statement

```
a .^ 2
```

results in

```
ans =
     1     4     9
    16    25    36
```

6.1.13 Matrices and `for`

If

```
a =
        1      2      3
        4      5      6
        7      8      9
```

the statements

```
for v = a
   disp(v')
end
```

result in

```
1      4      7
2      5      8
3      6      9
```

What happens in this most general form of the `for` statement is that the index v takes on the value of each *column* of the matrix expression a in turn. This provides a neat way of processing all the columns of a matrix. You can do the same with the rows if you transpose a, e.g., the statements

```
for v = a'
   disp(v')
end
```

display the rows of a one at a time. Get it?

6.1.14 Visualization of matrices

Matrices can be visualized graphically in MATLAB. This subject is discussed briefly in Chapter 9, with illustrations.

6.1.15 Vectorizing nested `for`s: loan repayment tables

If a regular fixed payment P is made n times a year to repay a loan of amount A over a period of k years, where the nominal annual interest rate is r, P is given by

$$P = \frac{rA(1+r/n)^{nk}}{n[(1+r/n)^{nk} - 1]}.$$

We would like to generate a table of repayments for a loan of $1 000 over 15, 20 or 25 years, at interest rates that vary from 10% to 20% per annum, in steps of 1%. Since P is directly proportional to A in the above formula, the repayments of a loan of any amount can be found by simple proportion from such a table.

The conventional way of handling this is with 'nested' `for`s. The `fprintf` statements are necessary in order to get the output for each interest rate on the same line (see MATLAB Help):

```
A = 1000;                   % amount borrowed
n = 12;                     % number of payments per year

for r = 0.1 : 0.01 : 0.2
  fprintf( '%4.0f%', 100 * r );
  for k = 15 : 5 : 25
    temp = (1 + r/n) ^ (n*k);
    P = r * A * temp / (n * (temp - 1));
    fprintf( '%10.2f', P );
  end;
  fprintf( '\n' );          % new line
end;
```

Some sample output (with headings edited in):

```
rate %    15 yrs    20 yrs    25 yrs

  10      10.75      9.65      9.09
  11      11.37     10.32      9.80
  ...
  19      16.83     16.21     15.98
  20      17.56     16.99     16.78
```

However, we saw in Chapter 2 that `for` loops can often be vectorized, saving a lot of computing time (and also providing an interesting intellectual challenge!). The inner loop can easily be vectorized; the following code uses only one `for`:

```
...
for r = 0.1 : 0.01 : 0.2
  k = 15 : 5 : 25;
  temp = (1 + r/n) .^ (n*k);
  P = r * A * temp / n ./ (temp - 1);
  disp( [100 * r, P] );
end;
```

Note the use of the array operators.

The really tough challenge, however, is to vectorize the outer loop as well. We want a table with 11 rows and 3 columns. Start by assigning values to A and n (from the command line):

```
A = 1000;
n = 12;
```

Then generate a column vector for the interest rates:

```
r = [0.1:0.01:0.2]'
```

Now change this into a table with 3 columns each equal to r:

```
r = repmat(r, [1 3])
```

The matrix r should look like this:

```
0.10        0.10        0.10
0.11        0.11        0.11
...
0.19        0.19        0.19
0.20        0.20        0.20
```

Now do a similar thing for the repayment periods k. Generate the row vector

```
k = 15:5:25
```

and expand it into a table with 11 rows each equal to k:

```
k = repmat(k, [11 1])
```

This should give you

```
15    20    25
15    20    25
...
15    20    25
15    20    25
```

The formula for P is a little complicated, so let's do it in two steps:

```
temp = (1 + r/n) .^ (n * k);
P = r * A .* temp / n ./ (temp - 1)
```

Finally, you should get this for P:

```
10.75          9.65          9.09
11.37         10.32          9.80
 . . .
16.83         16.21         15.98
17.56         16.99         16.78
```

This works, because of the way the tables r and k have been constructed, and because the MATLAB array operations are performed element by element. For example, when the calculation is made for P in row 2 and column 1, the array operations pick out row 2 of r (all 0.11) and column 1 of k (all 15), giving the correct value for P (11.37).

The nested for way might be easier to program, but this way is certainly more interesting (and quicker to execute).

6.1.16 Multi-dimensional arrays

MATLAB arrays can have more than two dimensions. For example, suppose you create the matrix

```
a = [1:2; 3:4]
```

You can add a third dimension to a with

```
a(:,:,2) = [5:6; 7:8]
```

MATLAB responds with

```
a(:,:,1) =
     1     2
     3     4
a(:,:,2) =
     5     6
     7     8
```

It helps to think of the 3-D array a as a series of 'pages', with a matrix on each page. The third dimension of a numbers the pages. This is analogous to a spreadsheet with multiple sheets: each sheet contains a table (matrix).

You can get into hyperpages with higher dimensions if you like!

See help datatypes for a list of special multi-dimensional array functions.

6.2 MATRIX OPERATIONS

We have seen that *array* operations are performed element by element on matrices. However, *matrix* operations, which are fundamental to MATLAB, are defined differently in certain cases, in the mathematical sense.

Matrix addition and subtraction are defined in the same way as the equivalent array operations, i.e., element by element. Matrix multiplication, however, is quite different.

6.2.1 Matrix multiplication

Matrix *multiplication* is probably the most important matrix operation. It is used widely in such areas as network theory, solution of linear systems of equations, transformation of co-ordinate systems, and population modeling, to name but a very few. The rules for multiplying matrices look a little weird if you've never seen them before, but will be justified by the applications that follow.

When two matrices A and B are multiplied together in this sense, their product is a third matrix C. The operation is written as

$$C = AB,$$

and the general element c_{ij} of C is formed by taking the *scalar product* of the ith row of A with the jth column of B. (The scalar product of two *vectors* x and y is $x_1 y_1 + x_2 y_2 + \ldots$, where x_i and y_i are the components of the vectors.) It follows that A and B can only be successfully multiplied (in that order) if the number of columns in A is the same as the number of rows in B.

The general definition of matrix multiplication is as follows: If A is a $n \times m$ matrix and B is a $m \times p$ matrix, their product C will be a $n \times p$ matrix such that the general element c_{ij} of C is given by

$$c_{ij} = \sum_{k=1}^{m} a_{ik} b_{kj}.$$

Note that in general AB is not equal to BA (matrix multiplication is not *commutative*).

Example:

$$\begin{bmatrix} 1 & 2 \\ 3 & 4 \end{bmatrix} \times \begin{bmatrix} 5 & 6 \\ 0 & -1 \end{bmatrix} = \begin{bmatrix} 5 & 4 \\ 15 & 14 \end{bmatrix}$$

Since a vector is simply a one-dimensional matrix, the definition of matrix multiplication given above also applies when a vector is multiplied by an ap-

propriate matrix, e.g.,

$$\begin{bmatrix} 1 & 2 \\ 3 & 4 \end{bmatrix} \times \begin{bmatrix} 2 \\ 3 \end{bmatrix} = \begin{bmatrix} 8 \\ 18 \end{bmatrix}.$$

The operator * is used for matrix multiplication, as you may have guessed. For example, if

```
a =
     1       2
     3       4
```

and

```
b =
     5       6
     0      -1
```

the statement

```
c = a * b
```

results in

```
c =
      5       4
     15      14
```

Note the important difference between the *array* operation a .* b (evaluate by hand and check with MATLAB) and the *matrix* operation a * b.

To multiply a matrix by a vector in that order, the vector must be a column vector. So if

```
b = [2 3]'
```

the statement

```
c = a * b
```

results in

```
c =
      8
     18
```

6.2.2 Matrix exponentiation

The matrix operation $\mathbf{A}^2$ means $\mathbf{A} \times \mathbf{A}$, where $\mathbf{A}$ must be a square matrix. The operator ^ is used for matrix exponentiation, e.g., if

```
a =
      1      2
      3      4
```

the statement

```
a ^ 2
```

results in

```
ans =
      7     10
     15     22
```

which is the same as a * a (try it).

Again, note the difference between the array operation a .^ 2 (evaluate by hand and check with MATLAB) and the matrix operation a ^ 2.

6.3 OTHER MATRIX FUNCTIONS

Here are some of MATLAB's more advanced matrix functions.

det determinant.

eig eigenvalue decomposition.

expm matrix exponential, i.e., $e^{\mathbf{A}}$, where $\mathbf{A}$ is a matrix. The matrix exponential may be used to evaluate analytical solutions of linear ordinary differential equations with constant coefficients.

inv inverse.

lu LU factorization (into lower and upper triangular matrices).

qr orthogonal factorization.

svd singular value decomposition.

6.4 POPULATION GROWTH: LESLIE MATRICES

Our first application of matrices is in population dynamics.

Suppose we want to model the growth of a population of rabbits, in the sense that given their number at some moment, we would like to estimate the size of the population in a few years' time. One approach is to divide the rabbit population up into a number of age classes, where the members of each age class are one time unit older than the members of the previous class, the time unit being whatever is convenient for the population being studied (days, months, etc.).

If X_i is the size of the ith age class, we define a *survival factor* P_i as the proportion of the ith class that survive to the $(i + 1)$th age class, i.e. the proportion that 'graduate'. F_i is defined as the *mean fertility* of the ith class. This is the mean number of newborn individuals expected to be produced during one time interval by each member of the ith class at the beginning of the interval (only females count in biological modeling, since there are always enough males to go round!).

Suppose for our modified rabbit model we have three age classes, with X_1, X_2 and X_3 members, respectively. We will call them young, middle-aged and old-aged for convenience. We will take our time unit as one month, so X_1 is the number that were born during the current month, and which will be considered as youngsters at the end of the month. X_2 is the number of middle-aged rabbits at the end of the month, and X_3 the number of oldsters. Suppose the youngsters cannot reproduce, so that $F_1 = 0$. Suppose the fertility rate for middle-aged rabbits is 9, so $F_2 = 9$, while for oldsters $F_3 = 12$. The probability of survival from youth to middle-age is one third, so $P_1 = 1/3$, while no less than half the middle-aged rabbits live to become oldsters, so $P_2 = 0.5$ (we are assuming for the sake of illustration that all old-aged rabbits die at the end of the month—this can be corrected easily). With this information we can quite easily compute the changing population structure month by month, as long as we have the population breakdown to start with.

If we now denote the current month by t, and next month by $(t + 1)$, we can refer to this month's youngsters as $X_1(t)$, and to next month's as $X_1(t+1)$, with similar notation for the other two age classes. We can then write a scheme for updating the population from month t to month $(t + 1)$ as follows:

$$X_1(t + 1) = F_2 X_2(t) + F_3 X_3(t),$$
$$X_2(t + 1) = P_1 X_1(t),$$
$$X_3(t + 1) = P_2 X_2(t).$$

We now define a population vector $\mathbf{X}(t)$, with three components, $X_1(t)$, $X_2(t)$, and $X_3(t)$, representing the three age classes of the rabbit population in

month t. The above three equations can then be rewritten as

$$\begin{bmatrix} X_1 \\ X_2 \\ X_3 \end{bmatrix}_{(t+1)} = \begin{bmatrix} 0 & F_2 & F_3 \\ P_1 & 0 & 0 \\ 0 & P_2 & 0 \end{bmatrix} \times \begin{bmatrix} X_1 \\ X_2 \\ X_3 \end{bmatrix}_t$$

where the subscript at the bottom of the vectors indicates the month. We can write this even more concisely as the matrix equation

$$\mathbf{X}(t+1) = \mathbf{L}\,\mathbf{X}(t), \tag{6.1}$$

where $\mathbf{L}$ is the matrix

$$\begin{bmatrix} 0 & 9 & 12 \\ 1/3 & 0 & 0 \\ 0 & 1/2 & 0 \end{bmatrix}$$

in this particular case. $\mathbf{L}$ is called a *Leslie matrix*. A population model can always be written in the form of Equation (6.1) if the concepts of age classes, fertility, and survival factors, as outlined above, are used.

Now that we have established a matrix representation for our model, we can easily write a script using matrix multiplication and repeated application of Equation (6.1):

$$\mathbf{X}(t+2) = \mathbf{L}\,\mathbf{X}(t+1),$$
$$\mathbf{X}(t+3) = \mathbf{L}\,\mathbf{X}(t+2), \text{ etc.}$$

We will assume to start with that we have one old (female) rabbit, and no others, so $X_1 = X_2 = 0$, and $X_3 = 1$. Here is the script:

```
% Leslie matrix population model
n = 3;
L = zeros(n);           % all elements set to zero
L(1,2) = 9;
L(1,3) = 12;
L(2,1) = 1/3;
L(3,2) = 0.5;
x = [0 0 1]';           % remember x must be a column vector!

for t = 1:24
  x = L * x;
  disp( [t x' sum(x)] )  % x' is a row
end
```

The output, over a period of 24 months (after some editing), is:

FIGURE 6.1
Total rabbit population over 15 months

Month	Young	Middle	Old	Total
1	12	0	0	12
2	0	4	0	4
3	36	0	2	38
4	24	12	0	36
5	108	8	6	122
...				
22	11184720	1864164	466020	13514904
23	22369716	3728240	932082	27030038
24	44739144	7456572	1864120	54059836

It so happens that there are no 'fractional' rabbits in this example. If there are any, they should be kept, and not rounded (and certainly not truncated). They occur because the fertility rates and survival probabilities are averages.

If you look carefully at the output you may spot that after some months the total population doubles every month. This factor is called the *growth factor*, and is a property of the particular Leslie matrix being used (for those who know about such things, it's the *dominant eigenvalue* of the matrix). The growth factor is 2 in this example, but if the values in the Leslie matrix are changed, the long-term growth factor changes too (try it and see).

Fig. 6.1 shows how the total rabbit population grows over the first 15 months. To draw this graph yourself, insert the line

```
p(t) = sum(x);
```

in the `for` loop after the statement $x = L * x$;, and run the program again. The vector p will contain the total population at the end of each month. Then enter the commands

```
plot(1:15, p(1:15)), xlabel('months'), ylabel('rabbits')
hold, plot(1:15, p(1:15),'o')
```

The graph demonstrates *exponential* growth. If you plot the population over the full 24-month period, you will see that the graph gets much steeper. This is a feature of exponential growth.

You probably didn't spot that the numbers in the three age classes tend to a limiting ratio of 24:4:1. This can be demonstrated very clearly if you run the model with an initial population structure having this limiting ratio. The limiting ratio is called the *stable age distribution* of the population, and again it is a property of the Leslie matrix (in fact, it is the *eigenvector* belonging to the dominant eigenvalue of the matrix). Different population matrices lead to different stable age distributions.

The interesting point about this is that a given Leslie matrix always eventually gets a population into the *same* stable age distribution, which increases eventually by the *same* growth factor each month, *no matter what the initial population breakdown is*. For example, if you run the above model with any other initial population, it will always eventually get into a stable age distribution of 24:4:1 with a growth factor of 2 (try it and see).

See `help eig` if you're interested in using MATLAB to compute eigenvalues and eigenvectors.

6.5 MARKOV PROCESSES

Often a process that we wish to model may be represented by a number of possible *discrete* (i.e. discontinuous) states that describe the outcome of the process. For example, if we are spinning a coin, then the outcome is adequately represented by the two states 'heads' and 'tails' (and nothing in between). If the process is random, as it is with spinning coins, there is a certain probability of being in any of the states at a given moment, and also a probability of changing from one state to another. If the probability of moving from one state to another depends on the present state only, and not on any previous state, the process is called a *Markov chain*. The progress of the myopic sailor in Chapter 13 is an example of such a process. Markov chains are used widely in such diverse fields as biology and business decision making, to name just two areas.

6.5.1 A random walk

This example is a variation on the random walk simulation of Chapter 13. A street has six intersections. A short-sighted student wanders down the street. His home is at intersection 1, and his favorite internet cafe at intersection 6. At each intersection other than his home or the cafe he moves in the direction of the cafe with probability 2/3, and in the direction of his home with probability 1/3. In other words, he is twice as likely to move towards the cafe as towards his home. He never wanders down a side street. If he reaches his home or the cafe, he disappears into them, never to re-appear (when he disappears we say in Markov jargon that he has been *absorbed*).

We would like to know: what are the chances of him ending up at home or in the cafe, if he starts at a given corner (other than home or the cafe, obviously)? He can clearly be in one of six states, with respect to his random walk, which can be labeled by the intersection number, where state 1 means *Home* and state 6 means *Cafe*. We can represent the probabilities of being in these states by a six-component *state vector* $\mathbf{X}(t)$, where $X_i(t)$ is the probability of him being at intersection i at moment t. The components of $\mathbf{X}(t)$ must sum to 1, since he has to be in one of these states.

We can express this Markov process with the following *transition probability matrix*, $\mathbf{P}$, where the rows represent the next state (i.e. corner), and the columns represent the present state:

	Home	2	3	4	5	Cafe
Home	1	1/3	0	0	0	0
2	0	0	1/3	0	0	0
3	0	2/3	0	1/3	0	0
4	0	0	2/3	0	1/3	0
5	0	0	0	2/3	0	0
Cafe	0	0	0	0	2/3	1

The entries for *Home-Home* and *Cafe-Cafe* are both 1 because he stays there with certainty.

Using the probability matrix $\mathbf{P}$ we can work out his chances of being, say, at intersection 3 at moment $(t + 1)$ as

$$X_3(t + 1) = 2/3X_2(t) + 1/3X_4(t).$$

To get to 3, he must have been at either 2 or 4, and his chances of moving from there are 2/3 and 1/3, respectively.

Mathematically, this is identical to the Leslie matrix problem. We can therefore form the new state vector from the old one each time with a matrix equation:

$$\mathbf{X}(t + 1) = \mathbf{P}\ \mathbf{X}(t).$$

If we suppose the student starts at intersection 2, the initial probabilities will be (0; 1; 0; 0; 0; 0). The Leslie matrix script may be adapted with very few changes to generate future states:

```
n = 6;
P = zeros(n);          % all elements set to zero

for i = 3:6
  P(i,i-1) = 2/3;
  P(i-2,i-1) = 1/3;
end

P(1,1) = 1;
P(6,6) = 1;
x = [0 1 0 0 0 0]';    % remember x must be a column vector!

for t = 1:50
  x = P * x;
  disp( [t x'] )
end
```

Edited output:

Time	Home	2	3	4	5	Cafe
1	0.3333	0	0.6667	0	0	0
2	0.3333	0.2222	0	0.4444	0	0
3	0.4074	0	0.2963	0	0.2963	0
4	0.4074	0.0988	0	0.2963	0	0.1975
5	0.4403	0	0.1646	0	0.1975	0.1975
6	0.4403	0.0549	0	0.1756	0	0.3292
7	0.4586	0	0.0951	0	0.1171	0.3292
8	0.4586	0.0317	0	0.1024	0	0.4073
...						
20	0.4829	0.0012	0	0.0040	0	0.5119
...						
40	0.4839	0.0000	0	0.0000	0	0.5161
...						
50	0.4839	0.0000	0	0.0000	0	0.5161

By running the program for long enough, we soon find the limiting probabilities: he ends up at home about 48% of the time, and at the cafe about 52% of the time. Perhaps this is a little surprising; from the transition probabilities, we might have expected him to get to the cafe rather more easily. It just goes to show that you should never trust your intuition when it comes to statistics!

Note that the Markov chain approach is *not* a simulation: one gets the *theoretical* probabilities each time (this can all be done mathematically, without a computer).

6.6 LINEAR EQUATIONS

A problem that often arises in scientific applications is the solution of a system of linear equations, e.g.,

$$3x + 2y - z = 10 \tag{6.2}$$
$$-x + 3y + 2z = 5 \tag{6.3}$$
$$x - y - z = -1. \tag{6.4}$$

MATLAB was designed to solve a system like this directly and very easily, as we shall now see.

If we define the matrix of coefficients, $\mathbf{A}$, as

$$\mathbf{A} = \begin{bmatrix} 3 & 2 & -1 \\ -1 & 3 & 2 \\ 1 & -1 & -1 \end{bmatrix},$$

and the vectors of unknowns, $\mathbf{x}$, and the right-hand side, $\mathbf{b}$, as

$$\mathbf{x} = \begin{bmatrix} x \\ y \\ z \end{bmatrix}, \qquad \mathbf{b} = \begin{bmatrix} 10 \\ 5 \\ -1 \end{bmatrix},$$

we can write the above system of three equations in matrix form as

$$\begin{bmatrix} 3 & 2 & -1 \\ -1 & 3 & 2 \\ 1 & -1 & -1 \end{bmatrix} \begin{bmatrix} x \\ y \\ z \end{bmatrix} = \begin{bmatrix} 10 \\ 5 \\ -1 \end{bmatrix},$$

or even more concisely as the single matrix equation

$$\mathbf{A}\mathbf{x} = \mathbf{b}. \tag{6.5}$$

The solution may then be written as

$$\mathbf{x} = \mathbf{A}^{-1}\mathbf{b}, \tag{6.6}$$

where $\mathbf{A}^{-1}$ is the matrix inverse of $\mathbf{A}$ (i.e., the matrix which when multiplied by $\mathbf{A}$ gives the identity matrix $\mathbf{I}$).

6.6.1 MATLAB's solution

To see how MATLAB solves this system, first recall that the *left division* operator \ may be used on scalars, i.e., a \ b is the same as b / a if a and b are scalars. However, it can also be used on vectors and matrices, in order to solve linear equations. Enter the following statements on the command line to solve Equations (6.2)–(6.4):

```
A = [3 2 -1; -1 3 2; 1  -1  -1];
b = [10 5 -1]';
x = A \ b
```

This should result in

```
x =
   -2.0000
    5.0000
   -6.0000
```

In terms of our notation, this means that the solution is $x = -2$, $y = 5$, $z = -6$.

You can think of the matrix operation A \ b as 'b divided by A', or as 'the inverse of A multiplied by b', which is essentially what Equation (6.6) means. A colleague of mine reads this operation as 'A under b', which you may also find helpful.

You may be tempted to implement Equation (6.6) in MATLAB as follows:

```
x = inv(A) * b
```

since the function inv finds a matrix inverse directly. However, A \ b is actually more accurate and efficient; see **MATLAB Help: Functions — Alphabetical List** and click on inv.

6.6.2 The residual

Whenever we solve a system of linear equations numerically we need to have some idea of how accurate the solution is. The first thing to check is the *residual*, defined as

```
r = A*x - b
```

where x is the result of the operation x = A \ b. Theoretically the residual r should be zero, since the expression A * x is supposed to be equal to b, according to Equation (6.5), which is the equation we are trying to solve. In our example here the residual is (check it)

```
r =
   1.0e-014 *
          0
    -0.3553
          0
```

This seems pretty conclusive: all the elements of the residual are less than 10^{-14} in absolute value. Unfortunately, there may still be problems lurking beneath the surface, as we shall see shortly.

We will, however, first look at a situation where the residual turns out to be far from zero.

6.6.3 Over-determined systems

When we have more equations than unknowns, the system is called *over-determined*, e.g.,

$$x - y = 0$$
$$y = 2$$
$$x = 1.$$

Surprisingly, perhaps, MATLAB gives a solution to such a system. If

```
A = [1 -1; 0 1; 1 0];
b = [0 2 1]';
```

the statement

```
x = A \ b
```

results in

```
x =
     1.3333
     1.6667
```

The residual `r = A*x - b` is now

```
r =
    -0.3333
    -0.3333
     0.3333
```

What happens in this case is that MATLAB produces the *least squares best fit*. This is the value of x which makes the *magnitude* of r, i.e.,

$$\sqrt{r(1)^2 + r(2)^2 + r(3)^3},$$

as small as possible. You can compute this quantity (0.5774) with `sqrt(r' * r)` or `sqrt(sum(r .* r))`. There is a nice example of fitting a decaying exponential function to data with a least squares fit in **MATLAB Help: Mathematics: Matrices and Linear Algebra: Solving Linear Equations: Overdetermined Systems**.

6.6.4 Under-determined systems

If there are fewer equations than unknowns, the system is called *under-determined*. In this case there are an infinite number of solutions; MATLAB will find one which has zeros for some of the unknowns.

The equations in such a system are the constraints in a linear programming problem.

6.6.5 Ill conditioning

Sometimes the coefficients of a system of equations are the results of an experiment, and may be subject to error. We need in that case to know how sensitive the solution is to the experimental errors. As an example, consider the system

$$10x + 7y + 8z + 7w = 32$$
$$7x + 5y + 6z + 5w = 23$$
$$8x + 6y + 10z + 9w = 33$$
$$7x + 5y + 9z + 10w = 31$$

Use matrix left division to show that the solution is $x = y = z = w = 1$. The residual is exactly zero (check it), and all seems well.

However, if we change the right-hand side constants to 32.1, 22.9, 32.9 and 31.1, the 'solution' is now given by $x = 6, y = -7.2, z = 2.9, w = -0.1$. The residual is very small.

A system like this is called *ill-conditioned*, meaning that a small change in the coefficients leads to a large change in the solution. The MATLAB function `rcond` returns the *condition estimator*, which tests for ill conditioning. If A is the coefficient matrix, `rcond(A)` will be close to zero if A is ill-conditioned, but close to 1 if it is well-conditioned. In this example, the condition estimator is about 2×10^{-4}, which is pretty close to zero.

Some authors suggest the rule of thumb that a matrix is ill-conditioned if its determinant is small compared to the entries in the matrix. In this case the

determinant of A is 1 (check with the function det) which is about an order of magnitude smaller than most of its entries.

6.6.6 Matrix division

Matrix left division, A\B, is defined whenever B has as many rows as A. This corresponds formally to inv(A)*B although the result is obtained without computing the inverse explicitly. In general,

```
x = A \ B
```

is a solution to the system of equations defined by $\mathbf{Ax} = \mathbf{B}$.

If A is square, matrix left division is done using Gauss elimination.

If A is not square the over- or under-determined equations are solved in the least squares sense. The result is a $m \times n$ matrix X, where m is the number of columns of A and n is the number of columns of B.

Matrix right division, B/A, is defined in terms of matrix left division such that B/A is the same as (A'\B')'. So with a and b defined as for Equations (6.2)–(6.4), this means that

```
x = (b'/a')'
```

gives the same solution, doesn't it? Try it, and make sure you can see why.

Sometimes the least squares solutions computed by \ or / for over- or under-determined systems can cause surprises, since you can legally divide one *vector* by another. For example, if

```
a = [1 2];
b = [3 4];
```

the statement

```
a / b
```

results in

```
ans =
    0.4400
```

This is because a/b is the same as (b'\a')', which is formally the solution of $\mathbf{b}'x' = \mathbf{a}'$. The result is a scalar, since a' and b' each have one column, and is

the least squares solution of

$$\begin{pmatrix} 3 \\ 4 \end{pmatrix} x = \begin{pmatrix} 1 \\ 2 \end{pmatrix}.$$

With this under your belt, can you explain why

```
a \ b
```

gives

```
ans =
        0        0
   1.5000   2.0000
```

(try writing the equations out in full)?

A complete discussion of the algorithms used in solving simultaneous linear equations may be found in the online documentation under **MATLAB: Reference: MATLAB Function Reference: Alphabetical List of Functions: Arithmetic Operators** + - * / \ ^ '.

6.7 SPARSE MATRICES

Matrices can sometimes get very large, with thousands of entries. Very large matrices can occupy huge amounts of memory, and processing them can take up a lot of computer time. For example, a system of n simultaneous linear equations requires n^2 matrix entries, and the computing time to solve them is proportional to n^3.

However, some matrices have relatively few *non-zero* entries. Such matrices are called *sparse* as opposed to *full*. The father of modern numerical linear algebra, J.H. Wilkinson, has remarked that a matrix is sparse if 'it has enough zeros that it pays to take advantage of them'. MATLAB has facilities for exploiting sparsity, which have the potential for saving huge amounts of memory and processing time. For example, the matrix representation of a certain type of partial differential equation (a 5-point Laplacian) on a square 64 × 64 grid is a 4096 × 4096 element matrix with 20224 non-zero elements. The sparse form of this matrix in MATLAB occupies only 250 kB of memory, whereas the full version of the same matrix would occupy 128 MB, which is way beyond the limits of most desktop computers. The solution of the system **Ax = b** using sparse techniques is about 4000 times faster than solving the full case, i.e., 10 seconds instead of 12 hours!

In this section (which you can safely skip, if you want to) we will look briefly at how to create sparse matrices in MATLAB. For a full description of sparse matrix techniques consult **MATLAB Help: Mathematics: Sparse Matrices**.

First an example, then an explanation. The transition probability matrix for the random walk problem in Section 6.5 is a good candidate for sparse representation. Only ten of its 36 elements are non-zero. Since the non-zeros appear only on the diagonal and the sub- and super-diagonals, a matrix representing more intersections would be even sparser. For example, a 100×100 representation of the same problem would have only 198 non-zero entries, i.e., 1.98%.

To represent a sparse matrix all that MATLAB needs to record are the non-zero entries with their row and column indices. This is done with the sparse function. The transition matrix of Section 6.5 can be set up as a sparse matrix with the statements

```
n = 6;
P = sparse(1, 1, 1, n, n);
P = P + sparse(n, n, 1, n, n);
P = P + sparse(1:n-2, 2:n-1, 1/3, n, n);
P = P + sparse(3:n, 2:n-1, 2/3, n, n)
```

which result in (with format rat)

```
P =

    (1,1)        1
    (1,2)       1/3
    (3,2)       2/3
    (2,3)       1/3
    (4,3)       2/3
    (3,4)       1/3
    (5,4)       2/3
    (4,5)       1/3
    (6,5)       2/3
    (6,6)        1
```

Each line of the display of a sparse matrix gives a non-zero entry with its row and column, e.g., 2/3 in row 3 and column 2. To display a sparse matrix in full form, use the function

```
full(P)
```

which results in

```
ans =
         1            1/3           0            0            0            0
         0            0            1/3          0            0            0
         0            2/3           0            1/3          0            0
         0            0            2/3          0            1/3          0
         0            0            0            2/3          0            0
         0            0            0            0            2/3          1
```

(also with `format rat`).

The form of the `sparse` function used here is

```
sparse(rows, cols, entries, m, n)
```

This generates an m × n sparse matrix with non-zero `entries` having subscripts (`rows`, `cols`) (which may be vectors), e.g., the statement

```
sparse(1:n-2, 2:n-1, 1/3, n, n);
```

(with n = 6) creates a 6 × 6 sparse matrix with 4 non-zero elements, being 1/3's in rows 1–4 and columns 2–5 (most of the super-diagonal). Note that repeated use of `sparse` produces a number of 6 × 6 matrices, which must be added together to give the final form. Sparsity is therefore preserved by operations on sparse matrices. See `help` for more details of `sparse`.

It's quite easy to test the efficiency of sparse matrices. Construct a (full) identity matrix

```
a = eye(1000);
```

Determine the time it takes to compute a^2. Now let us take advantage of the sparseness of a. It is an ideal candidate for being represented as a sparse matrix, since only 1000 of its one million elements are non-zero. It is represented in sparse form as

```
s = sparse(1:1000, 1:1000, 1, 1000, 1000);
```

Now check the time it takes to find a^2. Use `tic` and `toc` to find out how much faster is the computation.

The function `full(a)` returns the full form of the sparse matrix a (without changing the sparse representation of a itself). Conversely, `sparse(a)` returns the sparse form of the full matrix a.

The function `spy` provides a neat visualization of sparse matrices. Try it on P. Then enlarge P to about 50 × 50, and `spy` it.

SUMMARY

- A matrix is a 2-D array. Elements may be referenced in the conventional way with two subscripts. Alternatively, one subscript may be used, in which case the matrix is 'unwound' by columns.
- The colon operator may be used to represent all the rows or columns of a matrix, and also as a single subscript.
- The keyword `end` refers to the last row or column of a matrix.
- Use `repmat` to duplicate rows or columns of a matrix.
- Use the empty array `[ ]` to delete rows or columns of a matrix.
- Arrays may have more than two dimensions. In the case of a 3-D array the third subscript may be thought of as numbering pages, where each page contains a matrix defined by the first two subscripts.
- The matrix operations of multiplication and exponentiation are implemented with the matrix operators `*` and `^`.
- The matrix left division operator `\` is used for solving systems of linear equations directly. Because the matrix division operators `\` and `/` can sometimes produce surprising results with the least squares solution method, you should always compute the residual when solving a system of equations.
- If you work with large matrices with relatively few non-zero entries you should consider using MATLAB's sparse matrix facilities.

EXERCISES

6.1 Set up any 3×3 matrix `a`. Write some command-line statements to perform the following operations on `a`:
 (a) interchange columns 2 and 3;
 (b) add a fourth column (of 0's);
 (c) insert a row of 1's as the new second row of `a` (i.e. move the current second and third rows down);
 (d) remove the second column.

6.2 Compute the limiting probabilities for the student in Section 6.5 when he starts at each of the remaining intersections in turn, and confirm that the closer he starts to the cafe, the more likely he is to end up there.
Compute `P^50` directly. Can you see the limiting probabilities in the first row?

6.3 Solve the equations

$$2x - y + z = 4$$
$$x + y + z = 3$$
$$3x - y - z = 1$$

using the left division operator. Check your solution by computing the residual. Also compute the determinant (det) and the condition estimator (rcond). What do you conclude?

6.4 This problem, suggested by R.V. Andree, demonstrates ill conditioning (where small changes in the coefficients cause large changes in the solution). Use the left division operator to show that the solution of the system

$$x + 5.000y = 17.0$$
$$1.5x + 7.501y = 25.503$$

is $x = 2$, $y = 3$. Compute the residual.

Now change the term on the right-hand side of the second equation to 25.501, a change of about one part in 12000, and find the new solution and the residual. The solution is completely different. Also try changing this term to 25.502, 25.504, etc. If the coefficients are subject to experimental errors, the solution is clearly meaningless. Use rcond to find the condition estimator and det to compute the determinant. Do these values confirm ill conditioning?

Another way to anticipate ill conditioning is to perform a *sensitivity analysis* on the coefficients: change them all in turn by the same small percentage, and observe what effect this has on the solution.

6.5 Use sparse to represent the Leslie matrix in Section 6.4. Test your representation by projecting the rabbit population over 24 months.

6.6 If you are familiar with *Gauss reduction* it is an excellent programming exercise to code a Gauss reduction directly with operations on the rows of the augmented coefficient matrix. See if you can write a function

```
x = mygauss(a, b)
```

to solve the general system $\mathbf{Ax} = \mathbf{b}$. Skillful use of the colon operator in the row operations can reduce the code to a few lines!

Test it on **A** and **b** with random entries, and on the systems in Section 6.6 and Exercise 16.4. Check your solutions with left division.

Function M-files

THE OBJECTIVE OF THIS CHAPTER IS TO ENABLE YOU TO:

■ Write your own function M-files.

CONTENTS

Example: Newton's method again ... 163

Basic rules 165
Subfunctions......... 170
Private functions.... 170
P-code files 170
Improving M-file performance with the Profiler 171

Function handles........... 171

Command/function duality 173

Function name resolution........ 173

Debugging M-files 174
Debugging a script . 174
Debugging a function 176

Recursion........ 176

Summary 177

Exercises 179

Supplementary material.......... 180

We have already seen that MATLAB has a number of *built-in* (compiled) functions, e.g., sin, sqrt, sum. You can verify that they are built-in by attempting to type them. Try type sin for example. MATLAB also has some functions in the form of *function M-files*, e.g., fzero, why. You can see what's in them with type, e.g., type why.

Functions are indispensable when it comes to breaking a problem down into manageable logical pieces. Hence, MATLAB enables you to create your own function M-files. A function M-file is similar to a script file in that it also has a .m extension. However, a function M-file differs from a script file in that a function M-file communicates with the MATLAB workspace only through specially designated *input* and *output arguments*.

We begin this chapter by extending the ideas about implementing function M-files introduced in Chapter 3, Section 3.2.2. We do this by example. We then deal with the basic rules and the various input/output possibilities available to you in the development of function M-files. For additional information and examples type help function in the Command Window and click on doc function at the bottom of the displayed information.

7.1 EXAMPLE: NEWTON'S METHOD AGAIN

Newton's method may be used to solve a general equation $f(x) = 0$ by repeating the assignment

$$x \text{ becomes } x - \frac{f(x)}{f'(x)},$$

where $f'(x)$ (i.e., df/dx) is the first derivative of $f(x)$. The process continues until successive approximations to x are close enough.

Suppose that $f(x) = x^3 + x - 3$, i.e., we want to solve the equation $x^3 + x - 3 = 0$ (another way of stating the problem is to say we want to find the *zero* of $f(x)$).

163

Essential MATLAB for Engineers and Scientists. https://doi.org/10.1016/B978-0-08-102997-8.00013-0

We have to be able to differentiate $f(x)$. That's quite easy here: $f'(x) = 3x^2 + 1$. We could write inline objects for both $f(x)$ and $f'(x)$, but for this example we will use function M-files instead.

Use the Editor to create and save (in the current MATLAB directory) the function file f.m as follows:

```
function y = f(x)
y = x^3 + x - 3;
```

Then create and save another function file df.m:

```
function y = df(x)
y = 3*x^2 + 1;
```

Now write a *separate* script file, newtgen.m (in the same directory), which will stop either when the *relative* error in x is less than 10^{-8}, or after 20 steps, say:

```
% Newton's method in general
% excludes zero roots!
steps = 0;                          % iteration counter
x = input( 'Initial guess: ' );     % estimate of root
re = 1e-8;                           % required relative error
myrel = 1;

while myrel > re & (steps < 20)
  xold = x;
  x = x - f(x)/df(x);
  steps = steps + 1;
  disp( [x f(x)] )
  myrel = abs((x-xold)/x);
end

if myrel <= re
  disp( 'Zero found at' )
  disp( x )
else
  disp( 'Zero NOT found' )
end
```

Note that there are *two* conditions that will stop the while loop: convergence, or the completion of 20 steps. Otherwise the script could run indefinitely.

Here is a sample run (with `format long e`), starting with $x = 1$:

```
Initial guess: 1
        1.250000000000000e+000      2.031250000000000e-001
        1.214285714285714e+000      4.737609329445558e-003
        1.213412175782825e+000      2.779086666571118e-006
        1.213411662762407e+000      9.583445148564351e-013
        1.213411662762230e+000     -4.440892098500626e-016

   Zero found at
        1.213411662762230e+000
```

Note:

- The variable `y` in the function files `f.m` and `df.m` is the *output argument*. It is a dummy variable, and defines how output will be sent back to the outside world.

By the way, you realize of course that you can use your own functions from the command line, don't you? For example,

> ≫ `f(2)`

should return 7 with `f.m` defined as above.

7.2 BASIC RULES

Try the following more general example, which returns the mean (`avg`) and standard deviation (`stdev`) of the values in the vector `x`. Although there are two MATLAB functions to do this (`mean` and `std`), it is useful to have them combined into one. Write a function file `stats.m`:

```
function [avg, stdev] = stats( x )          % function definition line
% STATS       Mean and standard deviation             % H1 line
%             Returns mean (avg) and standard          % Help text
%             deviation (stdev) of the data in the
%             vector x, using Matlab functions

avg = mean(x);                                      % function body
stdev = std(x);
```

Now test it in the Command Window with some random numbers, e.g.,

```
r = rand(100,1);
[a, s] = stats(r);
```

Note the following points about function M-files in general:

- **General form of a function**
 A function M-file *name*.m has the following general form:

  ```
  function [ outarg1, outarg2, ...] = name( inarg1, inarg2, ... )
  % comments to be displayed with help
  ...
  outarg1 = ... ;
  outarg2 = ... ;
  ...
  ```

- `function` **keyword**
 The function file *must* start with the keyword `function` (in the function definition line).

- **Input and output arguments**
 The input and output arguments (*inarg1*, *outarg1*, etc.) are 'dummy' variables, and serve only to define the function's means of communication with the workspace. Other variable names may therefore be used in their place when the function is called (referenced).
 You can think of the *actual* input arguments being copied into the dummy input arguments when the function is called. So when `stats(r)` is called in the above example, the actual input argument `r` is copied into the input argument `x` in the function file. When the function *returns* (i.e., execution of it is completed) the dummy output arguments `avg` and `stdev` in the function file are copied into the actual output arguments `a` and `s`.

- **Multiple output arguments**
 If there is more than one output argument, the output arguments *must* be separated by commas and enclosed in square brackets in the function definition line, as shown.
 However, when a function is called with more than one output argument, the actual output arguments may be separated by commas or spaces.
 If there is only one output argument square brackets are not necessary.

- **Naming convention for functions**
 Function names must follow the MATLAB rules for variable names.
 If the filename and the function definition line name are different, the internal name is ignored.

- **Help text**
 When you type `help` *function_name*, MATLAB displays the comment lines that appear between the function definition line and the first non-comment (executable or blank) line.

The first comment line is called the *H1 line*. The lookfor function searches on and displays only the H1 line. The H1 lines of all M-files in a directory are displayed under the **Description** column of the Desktop Current Directory browser.

You can make help text for an entire directory by creating a file with the name Contents.m that resides in the directory. This file must contain only comment lines. The contents of Contents.m is displayed with the command

> help *directory_name*

If a directory does not have a Contents.m file, this command displays the H1 line of each M-file in the directory.

- **Local variables: scope**

Any variables defined inside a function are inaccessible outside the function. Such variables are called *local variables*—they exist only inside the function, which has its own workspace separate from the base workspace of variables defined in the Command Window.

This means that if you use a variable as a loop index, say, inside a function, it will not clash with a variable of the same name in the workspace or in another function.

You can think of the *scope* of a variable as the range of lines over which the variable is accessible.

- **Global variables**

Variables which are defined in the base workspace are not normally accessible inside functions, i.e., their scope is restricted to the workspace itself, unless they have been declared global, e.g.,

> global PLINK PLONK

If several functions, and possibly the base workspace, all declare particular variables as global, then they all share single copies of those variables. MATLAB recommends that global variables be typed in capital letters, to remind you that they are global.

The function isglobal(A) returns 1 if A is global, and 0 otherwise.

The command who global gives a list of global variables.

Use clear global to make all variables non-global, or clear PLINK to make PLINK non-global.

- **Persistent variables**

A variable in a function may be declared persistent. Local variables normally cease to exist when a function returns. Persistent variables, however, remain in existence between function calls. A persistent variable is initialized to the empty array.

In the following example, the persistent variable count is used to count how many times the function test is called:

```
function test
persistent count
if isempty(count)
    count = 1
else
    count = count + 1
end
```

Persistent variables remain in memory until the M-file is cleared or changed, e.g.,

```
clear test
```

The function `mlock` inside an M-file prevents the M-file from being cleared. A locked M-file is unlocked with `munlock`. The function `mislocked` indicates whether an M-file can be cleared or not.

The Help entry on `persistent` declares confidently: 'It is an error to declare a variable persistent if a variable with the same name exists in the current workspace.' However, this is definitely not the case at the time of writing (I tried it!).

■ **Functions that do not return values**

You might want to write a function that doesn't return values (such functions are called *procedures* or *subroutines* in languages like Pascal and Fortran, and *void* in C++ and Java). In that case you simply omit the output argument(s) and the equal sign in the function definition line. For example, the following function will display n asterisks:

```
function stars(n)
asteriks = char(abs('*')*ones(1,n));
disp( asteriks )
```

Go back to Chapter 6 for an explanation of how it works.

Write such a function file (`stars.m`), and test it out, e.g., `stars(13)` should produce 13 asterisks.

■ **Vector arguments**

It should come as no surprise that input and output arguments may be vectors, e.g., the following function generates a vector of *n* random rolls of a die:

```
function d = dice( n )
d = floor( 6 * rand(1, n) + 1 );
```

When an output argument is a vector, it is initialized each time the function is called, any previous elements being cleared. Its size at any moment is therefore determined by the most recent call of the function. For example, suppose the function `test.m` is defined as

```
function a = test
a(3) = 92;
```

Then if b is defined in the base workspace as

```
b =
     1     2     3     4     5     6
```

the statement

```
b = test
```

results in

```
b =
     0     0    92
```

- **How function arguments are passed**
 If a function changes the value of any of its input arguments, the change is *not reflected* in the actual input argument on return to the workspace (unless the function is called with the same input and output argument—see below). For the technically minded, input arguments appear to be passed *by value.*
 You might think that passing a large matrix as an input argument by value is wasteful of memory, and you would be correct. However, the designers of MATLAB were aware of this, and so an input argument is only passed by value if a function modifies it (although the modification is not reflected on return). If a function does not modify an input argument it is passed by reference.
- **Simulated pass by reference**
 A function may be called with the same actual input and output argument. For example, the following function `prune.m` removes all the zero elements from its input argument:

```
function y = prune(x)
y = x(x ~= 0);
```

 (if you can't figure out why, refer to **Subscripting with logical vectors** in Chapter 5).
 You can use it to remove all the zero elements of the vector x as follows:

```
x = prune(x)
```

- **Checking the number of function arguments**
 A function may be called with all, some, or none of its input arguments. If called with no arguments, the parentheses must be omitted. You may not use more input arguments than appear in its definition.
 The same applies to output arguments—you may specify all, some, or none of them when you use the function. If you call a function with no output arguments, the value of the first one in the definition is returned.
 There are times when a function may need to know how many input/output arguments are used on a particular call. In that case, the functions `nargin` and `nargout` can be used to determine the number of actual input and output arguments. For example,

```
function y = foo(a, b, c);
disp( nargin );
...
```

will display the number of input arguments present on each call of `foo`.

■ **Passing a variable number of arguments**

The functions `varargin` and `varargout` allow you to call a function with any number of input or output arguments. Since this facility involves packing arguments into a *cell array*, discussion of it is deferred to Chapter 11.

7.2.1 Subfunctions

A function M-file may contain the code for more than one function. The first function in a file is the *primary* function, and is the one invoked with the M-file name. Additional functions in the file are called *subfunctions*, and are visible only to the primary function and to other subfunctions.

Each subfunction begins with its own function definition line. Subfunctions follow each other in any order *after* the primary function.

7.2.2 Private functions

A private function is a function residing in a subdirectory with the name `private`. Private functions are visible only to functions in the parent directory. See Help for more details.

7.2.3 P-code files

The first time a function is called during a MATLAB session it is parsed ('compiled') into pseudocode and stored in memory to obviate the need for it to be parsed again during the current session. The pseudocode remains in memory until you clear it with `clear` *function_name* (see Help for all the possibilities with `clear`).

You can use the `pcode` function to save the parsed version of an M-file for use in later MATLAB sessions, or by users from whom you want to hide your algorithms. For example, the command

```
pcode stats
```

parses `stats.m` and stores the resulting pseudocode in the file named `stats.p`.

MATLAB is very efficient at parsing so producing your own P-code files seldom makes much of a speed difference, except in the case of large GUI applications where many M-files have to be parsed before the application surfaces.

7.2.4 Improving M-file performance with the Profiler

The MATLAB Profiler enables you to see where the bottlenecks in your programs are, e.g., which functions are consuming most of the time. With this information you can often redesign programs to be more efficient. To find out more about this utility open the **MATLAB Help** documentation via the '?' at the top of the desktop and type "Profiler" in the search space. You should be at the document titled "The Profiler Utility".

7.3 FUNCTION HANDLES

Our script file `newtgen.m` in Section 7.1 solves the equation $f(x) = 0$, where $f(x)$ is defined in the function file with the specific name `f.m`. This is restrictive because to solve a different equation `f.m` has to be edited first. To make `newtgen` even more general, it can be rewritten as a function M-file itself, with a *handle* to `f.m` as an input argument. This process is made possible by the built-in function `feval`, and the concept of a function handle, which we now examine.

Try the following on the command line:

```
fhandle = @sqrt;
feval(fhandle, 9)
feval(fhandle, 25)
```

Can you see that `feval(fhandle, x)` is the same as `sqrt(x)`? The statement

```
fhandle = @sqrt
```

creates a *handle* to the function `sqrt`. The handle provides a way of referring to the function, for example in a list of input arguments to another function. A MATLAB function handle is similar to a pointer in C++, although more general.

If you have still got a function file `f.m` defined for $f(x) = x^3 + x - 3$, verify that

```
feval(@f, 2)
```

for example, returns the same value as `f(2)`.

- In general, the first argument of `feval` is a handle to the function to be evaluated in terms of the subsequent arguments of `feval`.

You can use `feval` inside a function to evaluate another function whose handle is passed as an argument. as we will see now. As an example, we would like to rewrite our `newtgen` script as a function `newtfun`, to be called as follows,

```
[x f conv] = newtfun( fh, dfh, x0 )
```

where `fh` and `dfh` are handles for the M-files containing $f(x)$ and $f'(x)$ respectively, and `x0` is the initial guess. The outputs are the zero, the function value at the zero, and an argument `conv` to indicate whether or not the process has converged. The complete M-file `newtfun.m` is as follows:

```
function [x, f, conv] = newtfun(fh, dfh, x0)
% NEWTON    Uses Newton's method to solve f(x) = 0.
%           fh is handle to f(x), dfh is handle to f'(x).
%           Initial guess is x0.
%           Returns final value of x, f(x), and
%           conv (1 = convergence, 0 = divergence)

steps = 0;                          % iteration counter
x = x0;
re = 1e-8;                          % required relative error
myrel = 1;

while myrel > re & (steps < 20)
    xold = x;
    x = x - feval(fh, x)/feval(dfh, x);
    steps = steps + 1;
    disp( [x feval(fh, x)] )
    myrel = abs((x-xold)/x);
end;

if myrel <= re
    conv = 1;
else
    conv = 0;
end;

f = feval(fh, x);
```

Try it out. Verify that you can call `newtfun` with less than three output variables. Also check `help newton`.

A function handle gives you more than just a reference to the function. Open the help document and search for "function handles" to find more information on this topic.

Functions, such as `feval`, `fplot`, `newtfun`, etc. which take function handles as arguments are referred to by MATLAB as *function functions*, as opposed to functions which take numeric arrays as arguments.

Use of a function handle to evaluate a function supersedes the earlier use of `feval` where a string containing the function's name was passed as an argument.

7.4 COMMAND/FUNCTION DUALITY

In the earliest versions of MATLAB there was a clear distinction between *commands* like

```
clear
save junk x y z
whos
```

and *functions* like

```
sin(x)
plot(x, y)
```

If commands had any arguments they had to be separated by blanks with no brackets. Commands altered the environment, but didn't return results. New commands could not be created with M-files.

From Version 4 onwards commands and functions are 'duals', in that commands are considered to be functions taking string arguments. So

```
axis off
```

is the same as

```
axis( 'off' )
```

Other examples are

```
disp Error!
hold('on')
```

This duality makes it possible to generate command arguments with string manipulations, and also to create new commands with M-files.

7.5 FUNCTION NAME RESOLUTION

Remember that a variable in the workspace can 'hide' a built-in function of the same name, and a built-in function can hide an M-file.

Specifically, when MATLAB encounters a name it resolves it in the following steps:

1. Checks if the name is a variable.
2. Checks if the name is a *subfunction* of the calling function.
3. Checks if the name is a *private function*.
4. Checks if the name is in the directories specified by MATLAB's search path.

MATLAB therefore always tries to use a name as a variable first, before trying to use it as a script or function.

7.6 DEBUGGING M-FILES

Run-time errors (as opposed to syntax errors) which occur inside function M-files are often hard to fix, because the function workspace is lost when the error forces a return to the base workspace. The Editor/Debugger enables you to get inside a function while it is running, to see what's going wrong.

7.6.1 Debugging a script

To see how to debug interactively, let's first try the script `newtgen.m` in Section 7.3. Go through the following steps.

- Open `newtgen.m` with the MATLAB Editor/Debugger. (Incidentally, have you found out that you can run the Editor directly from the command line, e.g., with `edit newtgen`).
 You will have noticed that the lines in the Editor window are numbered. You can generate these line numbers from the command line for reference purposes with the command `dbtype`:

```
dbtype newtgen
```

```
1       % Newton's method in general
2       % exclude zero roots!
3
4       steps = 0;                      % iteration counter
5       x = input( 'Initial guess: ' ); % estimate of root
6       re = 1e-8;                      % required relative error
7       myrel = 1;
8
9       while myrel > re & (steps < 20)
        xold = x;
11        x = x - f(x)/df(x);
12        steps = steps + 1;
13        disp( [x f(x)] )
14        myrel = abs((x-xold)/x);
15      end;
```

```
16
17    if myrel <= re
18      disp( 'Zero found at' )
19      disp( x )
20    else
21      disp( 'Zero NOT found' )
22    end;
```

- To get into Debug mode you need to set a *breakpoint* just before where you think the problem might be. Alternatively, if you just want to 'step through' a script line by line, set a breakpoint at the first executable statement. The column to the right of the line numbers is called the *breakpoint alley*. You can only set breakpoints at executable statements—these are indicated by dashes in the breakpoint alley.

 Set a breakpoint at line 4 (steps = 0;) by clicking in the breakpoint alley. You can remove a breakpoint by clicking on the breakpoint icon, or using the Editor's **Breakpoints** menu (this menu also allows you to specify stopping conditions). You can also set/clear breakpoints on the current line with the set/clear breakpoint button on the toolbar.

- Having set your breakpoints, run the script in the Editor by clicking the run button in the toolbar, or with **Debug -> Run (F5)**.

 You can also run the script from the command line.

- When the script starts to run, there are two things in particular to notice. First, the symbol K appears to left of the command-line prompt to remind you that MATLAB is in debug mode.

 Second, a green arrow appears just to the right of the breakpoint in the Editor. The arrow indicates the next statement which is *about to be* executed.

- Now step through the script with **Debug -> Step (F10)**. Note when line 5 is executed you need to enter the value of x at the command line.

- When you get to line 11 (x = x - f(x)/df(x);) in this way, use **Debug -> Step In (F11)** and the Debugger will take you into the functions f.m and df.m.

- Continue with **F10**. Note that output appears in the Command Window as each disp statement is executed.

- There are a number of ways of examining variable values in debug mode:

 1. Position the cursor to the left of the variable in the Editor. Its current value appears in a box—this is called a datatip.
 2. Type the name of the variable in the Command Window.
 3. Use the Array Editor: open the Workspace browser and double-click a variable. If you arrange your windows carefully you can watch the value of a variable change in the Array Editor while you step through a program.

 Note that you can view variables only in the current workspace. The Editor has a **Stack** field to the right of the toolbar where you

can select the workspace. For example, if you have stepped into `f.m` the current workspace is shown as `f`. At this point you can view variables in the `newtgen` workspace by selecting `newtgen` in the **Stack** field.

You can use the Array Editor or the command line to change the value of a variable. You can then continue to see how the script performs with the new value.

- Another useful debugging feature is **Debug -> Go Until Cursor**. This enables you to continue running the script to the line where you've positioned the cursor.
- To quit debugging click the exit debug mode button in the Editor/Debugger toolbar, or select **Debug -> Exit Debug Mode**.
 If you forget to quit debugging you won't be able to get rid of the `K` prompt on the command line!

7.6.2 Debugging a function

You can't run a function directly in the Editor/Debugger—you have to set a breakpoint in the function and run it from the command line. Let's use `newtfun.m` as an example.

- Open `newtfun.m` in the Editor/Debugger.
- Set a breakpoint at line 8 (`steps = 0;`).
- In the Command Window set up function handles for `f` and `df` and call `newtfun`:

```
fhand = @f;
dfhand = @df;
[x f conv] = newtfun(fhand, dfhand, 10)
```

- Note that MATLAB goes into debug mode and takes you to the breakpoint in `newtfun`. Now you can continue debugging as before.

Debugging may also be done from the command line with the debug functions. See `help debug`.

7.7 RECURSION

Many (mathematical) functions are defined *recursively*, i.e., in terms of simpler cases of themselves, e.g., the factorial function may be defined recursively as

$$n! = n \times (n-1)!$$

as long as 1! is defined as 1. MATLAB allows functions to call themselves; this process is called *recursion*. The factorial function may be written recursively in an M-file `fact.m` like this:

```
function y = fact(n)
% FACT       Recursive definition of n!

if n > 1
  y = n * fact(n-1);
else
  y = 1;
end;
```

Recursive functions are usually written in this way: an if statement handles the general recursive definition; the else part handles the special case ($n = 1$).

Although recursion appears deceptively simple, it is an advanced topic, as the following experiment demonstrates. Insert the statement disp(n) into the definition of fact immediately *above* the if statement, and run fact(5) from the command line. The effect is what you might expect: the integers 5 to 1 in descending order. Now move disp(n) to *below* the if statement, and see what happens. The result is the integers 1 to 5 in *ascending* order now, which is rather surprising.

In the first case, the value of n is displayed each time fact is called, and the output is obvious enough. However, there is the world of difference between a recursive function being *called*, and *executed*. In the second case, the disp statement is only executed after the if has finished executing. And when is that exactly? Well, when the initial call to fact takes place, n has the value 5, so the first statement in the if is invoked. However, the value of fact(4) is not known at this stage, so a *copy* is made of all the statements in the function which will need to be executed once the value of fact(4) is known. The reference to fact(4) makes fact call itself, this time with a value of 4 for n. Again, the first statement in the if is invoked, and MATLAB discovers that it doesn't know the value of fact(3) this time. So another (different) copy is made of all the statements that will have to be executed once the value of fact(3) is known. And so each time fact is called, separate copies are made of all the statements *yet to be executed*. Finally, MATLAB joyfully finds a value of n (1) for which it actually knows the value of fact, so it can at last begin to execute (in reverse order) the pile of statements which have been damming up inside the memory.

This discussion illustrates the point that recursion should be treated with respect. While it is perfectly in order to use it in an example like this, it can chew up huge amounts of computer memory and time.

SUMMARY

- Good structured programming requires real problem-solving programs to be broken down into function M-files.

- The name of a function in the function definition line should be the same as the name of the M-file under which it is saved. The M-file must have the extension .m.
- A function may have input and output arguments, which are usually its only way of communicating with the workspace. Input/output arguments are dummy variables (placeholders).
- Comment lines up to the first non-comment line in a function are displayed when help is requested for the function.
- Variables defined inside a function are local variables and are inaccessible outside the function.
- Variables in the workspace are inaccessible inside a function unless they have been declared global.
- A function does not have to have any output arguments.
- Input arguments have the appearance of being passed by value to a function. This means that changes made to an input argument inside a function are not reflected in the actual input argument when the function returns.
- A function may be called with fewer than its full number of input/output arguments.
- The functions nargin and nargout indicate how many input and output arguments are used on a particular function call.
- Variables declared persistent inside a function retain their values between calls to the function.
- Subfunctions in an M-file are accessible only to the primary function and to other subfunctions in the same M-file.
- Private functions are functions residing in a sub-directory named private and are accessible only to functions in the parent directory.
- Functions may be parsed (compiled) with the pcode function. The resulting code has the extension .p and is called a P-code file.
- The Profiler enables you to find out where your programs spend most of their time.
- A handle for a function is created with @.
 A function may be represented by its handle. In particular the handle may be passed as an argument to another function.
- feval evaluates a function whose handle is passed to it as an argument.
- MATLAB first tries to use a name as a variable, then as a built-in function, and finally as one of the various types of function.
- Command/function duality means that new commands can be created with function M-files, and that command arguments may be generated with string manipulations.
- The Editor/Debugger enables you to work through a script or function line-by-line in debug mode, examining and changing variables on the way.
- A function may call itself. This feature is called recursion.

EXERCISES

7.1 Change the function `stars` of Section 7.2 to a function `pretty` so that it will draw a line of any specified character. The character to be used must be passed as an additional input (string) argument, e.g., `pretty(6, '$')` should draw six dollar symbols.

7.2 Write a script `newquot.m` which uses the Newton quotient $[f(x+h) - f(x)]/h$ to estimate the first derivative of $f(x) = x^3$ at $x = 1$, using successively smaller values of h: 1, 10^{-1}, 10^{-2}, etc. Use a function M-file for $f(x)$.

Rewrite `newquot` as a function M-file able to take a handle for $f(x)$ as an input argument.

7.3 Write and test a function `double(x)` which doubles its input argument, i.e., the statement `x = double(x)` should double the value in `x`.

7.4 Write and test a function `swop(x, y)` which will exchange the values of its two input arguments.

7.5 Write your own MATLAB function to compute the exponential function directly from the Taylor series:

$$e^x = 1 + x + \frac{x^2}{2!} + \frac{x^3}{3!} + \ldots$$

The series should end when the last term is less than 10^{-6}. Test your function against the built-in function `exp`, but be careful not to make x too large—this could cause rounding error.

7.6 If a random variable X is distributed normally with zero mean and unit standard deviation, the probability that $0 \leq X \leq x$ is given by the standard normal function $\Phi(x)$. This is usually looked up in tables, but it may be approximated as follows:

$$\Phi(x) = 0.5 - r(at + bt^2 + ct^3),$$

where $a = 0.4361836$, $b = -0.1201676$, $c = 0.937298$, $r = \exp(-0.5x^2)/\sqrt{2\pi}$, and $t = 1/(1 + 0.3326x)$.

Write a function to compute $\Phi(x)$, and use it in a program to write out its values for $0 \leq x \leq 4$ in steps of 0.1. Check: $\Phi(1) = 0.3413$.

7.7 Write a function

```
function [x1, x2, flag] = quad( a, b, c )
```

which computes the roots of the quadratic equation $ax^2 + bx + c = 0$. The input arguments `a`, `b` and `c` (which may take any values) are the coefficients of the quadratic, and `x1`, `x2` are the two roots (if they exist), which may be equal. See Fig. 3.3 in Chapter 3 for the structure plan. The output argument `flag` must return the following values, according to the number and type of roots:

0: no solution ($a = b = 0$, $c \neq 0$);
1: one real root ($a = 0$, $b \neq 0$, so the root is $-c/b$);
2: two real or complex roots (which could be equal if they are real);
99: any x is a solution ($a = b = c = 0$).

Test your function on the data in Exercise 3.5.

7.8 The Fibonacci numbers are generated by the sequence

$$1, 1, 2, 3, 5, 8, 13, \ldots$$

Can you work out what the next term is? Write a recursive function f(n) to compute the Fibonacci numbers F_0 to F_{20}, using the relationship

$$F_n = F_{n-1} + F_{n-2},$$

given that $F_0 = F_1 = 1$.

7.9 The first three Legendre polynomials are $P_0(x) = 1$, $P_1(x) = x$, and $P_2(x) = (3x^2 - 1)/2$. There is a general *recurrence* formula for Legendre polynomials, by which they are defined recursively:

$$(n + 1)P_{n+1}(x) - (2n + 1)x\,P_n(x) + n\,P_{n-1}(x) = 0.$$

Define a recursive function p(n,x) to generate Legendre polynomials, given the form of P_0 and P_1. Use your function to compute p(2,x) for a few values of x, and compare your results with those using the analytic form of $P_2(x)$ given above.

APPENDIX 7.A SUPPLEMENTARY MATERIAL

Supplementary material related to this chapter can be found online at https://doi.org/10.1016/B978-0-08-102997-8.00013-0.

Loops

THE OBJECTIVES OF THIS CHAPTER ARE TO ENABLE YOU TO LEARN TO:

- Program (or code) determinate loops with `for`.
- Program (or code) indeterminate loops with `while`.

CONTENTS

Determinate repetition with
for 181
Binomial coefficient 181
Update processes .. 182
Nested fors 184

Indeterminate repetition with
while 184
A guessing game ... 184
The while
statement............. 185
Doubling time of an investment 186
Prime numbers 187
Projectile trajectory 188
break *and* continue 190
Menus 191

Summary 192

Exercises 192

In Chapter 2 we introduced the powerful `for` statement, which is used to repeat a block of statements a fixed number of times. This type of structure, where the number of repetitions must be determined in advance, is sometimes called *determinate repetition*. However, it often happens that the condition to end a loop is only satisfied *during the execution of the loop itself*. Such a structure is called *indeterminate*. We begin with a discussion of determinate repetition.

8.1 DETERMINATE REPETITION WITH for

8.1.1 Binomial coefficient

The *binomial coefficient* is widely used in mathematics and statistics. It is defined as the number of ways of choosing r objects out of n without regard to order, and is given by

$$\binom{n}{r} = \frac{n!}{r!(n-r)!} .$$

$$(8.1)$$

If this form is used, the factorials can get very big, causing an overflow. But a little thought reveals we can simplify Equation (8.1) as follows:

$$\binom{n}{r} = \frac{n(n-1)(n-2)\cdots(n-r+1)}{r!} ,$$

$$(8.2)$$

$$\text{e.g.,} \quad \binom{10}{3} = \frac{10!}{3! \times 7!} = \frac{10 \times 9 \times 8}{1 \times 2 \times 3} .$$

Using Equation (8.2) is computationally much more efficient:

Essential MATLAB for Engineers and Scientists. https://doi.org/10.1016/B978-0-08-102997-8.00014-2

```
ncr = 1;
n = ...
r = ...

for k = 1:r
  ncr = ncr * (n - k + 1) / k;
end

disp( ncr )
```

The binomial coefficient is sometimes pronounced 'n-see-r'. Work through the program by hand with some sample values.

8.1.2 Update processes

Many problems in science and engineering involve modeling a process where the main variable is repeatedly updated over a period of time. Here is an example of such an *update process*.

A can of orange juice at temperature 25°C is placed in a fridge, where the ambient temperature F is 10°C. We want to find out how the temperature of the orange juice changes over a period of time. A standard way of approaching this type of problem is to break the time period up into a number of small steps, each of length dt. If T_i is the temperature at the *beginning* of step i, we can use the following model to get T_{i+1} from T_i:

$$T_{i+1} = T_i - K \, dt \, (T_i - F), \tag{8.3}$$

where K is a constant parameter depending on the insulating properties of the can, and the thermal properties of orange juice. Assume that units are chosen so that time is in minutes.

The following script implements this scheme. To make the solution more general, take the starting time as a, and the end time as b. If dt is very small, it will be inconvenient to have output displayed after every step, so the script also asks you for the *output interval* opint. This is the time (in minutes) between successive rows of output. It checks that this interval is an integer multiple of dt. Try the script with some sample values, e.g., $dt = 0.2$ minutes and opint $= 5$ minutes (these are the values used for the output below).

```
K = 0.05;
F = 10;
a = 0;              % start time
b = 100;            % end time
time = a;           % initialize time
T = 25;             % initialize temperature
load train          % prepare to blow the whistle
```

```
dt = input( 'dt: ' );
opint = input( 'output interval (minutes): ' );
if opint/dt ~= fix(opint/dt)
  sound(y, Fs)          % blow the whistle!
  disp( 'output interval is not a multiple of dt!' );
end

clc
format bank
disp( '          Time              Temperature' );
disp( [time T] )        % display initial values

for time = a+dt : dt : b
  T = T - K * dt * (T - F);
  if abs(rem(time, opint)) < 1e-6     % practically zero!
    disp( [time T] )
  end
end
```

Output:

```
    Time          Temperature
       0          25.00
    5.00          21.67
     ...
   95.00          10.13
  100.00          10.10
```

Note:

1. The function rem is used to display the results every opint minutes: when time is an integer multiple of opint its remainder when divided by opint should be zero. However, due to rounding error, the remainder is not always *exactly* zero. It is therefore better to test whether its absolute value is less than some very small value. (Rounding error is discussed in Chapter 9).

2. While this is probably the most obvious way of writing the script, we cannot easily plot the graph of temperature against time this way, since time and T are scalars which are repeatedly updated. To plot the graph they both need to be vectors (see Chapter 11).

3. Note how sound is implemented. See help audio for other interesting sounds supplied by MATLAB.

4. In case you are wondering how I got the headings in the right place, I'll let you into the secret. Run the script without a heading but with the numerical output as you want it. Then simply paste the `disp` statement with the headings into the command window and edit it until the headings fall in the right place. Paste the final version of the `disp` statement into the script.

5. Note the use of `break` to stop the script prematurely if the user gives bad input. See below for more general use of `break`.

8.1.3 Nested `for`s

As we saw in Chapter 6 (**Vectorizing nested `for`s**), `for` loops can be nested inside each other. The main point to note is that the index of the *inner* `for` moves faster.

8.2 INDETERMINATE REPETITION WITH `while`

Determinate loops all have in common the fact that you can work out in principle exactly how many repetitions are required *before the loop starts*. But in the next example, there is no way *in principle* of working out the number of repeats, so a different structure is needed.

8.2.1 A guessing game

The problem is easy to state. MATLAB 'thinks' of an integer between 1 and 10 (i.e., generates one at random). You have to guess it. If your guess is too high or too low, the script must say so. If your guess is correct, a message of congratulations must be displayed.

A little more thought is required here, so a structure plan might be helpful:

1. Generate random integer
2. Ask user for guess
3. While guess is wrong:
 If guess is too low
 Tell her it is too low
 Otherwise
 Tell her it is too high
 Ask user for new guess
4. Polite congratulations
5. Stop.

Here is the script:

```
matnum = floor(10 * rand + 1);
guess = input( 'Your guess please: ' );
load splat

while guess ~= matnum
sound(y, Fs)

  if guess > matnum
    disp( 'Too high' )
  else
    disp( 'Too low' )
  end;

  guess = input( 'Your next guess please: ' );
end

disp( 'At last!' )
load handel
sound(y, Fs)              % hallelujah!
```

Try it out a few times. Note that the while loop repeats as long as matnum is not equal to guess. There is no way in principle of knowing how many loops will be needed before the user guesses correctly. The problem is truly indeterminate.

Note that guess has to be input in two places: firstly to get the while loop going, and secondly during the execution of the while.

8.2.2 The while **statement**

In general the while statement looks like this:

```
while condition
    statements
end
```

The while construct repeats *statements* WHILE its *condition* remains true. The condition therefore is the condition to repeat once again. The condition is tested each time BEFORE *statements* are repeated. Since the condition is evaluated before *statements* are executed, it is possible to arrange for *statements* not to be executed at all under certain circumstances. Clearly, *condition* must depend on *statements* in some way, otherwise the loop will never end.

Recall that a *vector* condition is considered true only if *all* its elements are non-zero.

The command-line form of while is:

```
while condition statements, end
```

8.2.3 Doubling time of an investment

Suppose we have invested some money which draws 10 per cent interest per year, compounded. We would like to know how long it takes for the investment to double. More specifically, we want a statement of the account each year, *until* the balance has doubled. The English statement of the problem hints heavily that we should use an indeterminate loop with the following structure plan:

1. Initialize balance, year, interest rate
2. Display headings
3. Repeat
 Update balance according to interest rate
 Display year, balance
 until balance exceeds twice original balance
4. Stop.

A program to implement this plan would be

```
a = 1000;
r = 0.1;
bal = a;
year = 0;
disp( 'Year       Balance' )

while bal < 2 * a
   bal = bal + r * bal;
   year = year + 1;
   disp( [year bal] )
end
```

Note that the more natural phrase in the structure plan 'until balance exceeds twice original balance' must be coded as

```
while bal < 2 * a ...
```

This condition is checked each time before another loop is made. Repetition occurs only if the condition is true. Here's some output (for an opening balance of $1000):

```
Year       Balance

  1        1100.00
  2        1210.00
  3        1331.00
```

```
4          1464.10
5          1610.51
6          1771.56
7          1948.72
8          2143.59
```

Note that when the last loop has been completed, the condition to repeat is false for the first time, since the new balance ($2143.59) is more than $2000. Note also that a determinate `for` loop *cannot* be used here because we don't know how many loops are going to be needed until after the script has run (although in this particular example perhaps you *could* work out in advance how many repeats are needed?).

If you want to write the new balance only while it is *less* than $2000, all that has to be done is to move the statement

```
disp( [year bal] )
```

until it is the first statement in the `while` loop. Note that the initial balance of $1000 is displayed now.

8.2.4 Prime numbers

Many people are obsessed with prime numbers, and most books on programming have to include an algorithm to test if a given number is prime. So here's mine.

A number is prime if it is not an exact multiple of any other number except itself and 1, i.e., if it has no factors except itself and 1. The easiest plan of attack then is as follows. Suppose P is the number to be tested. See if any numbers N can be found that divide into P without remainder. If there are none, P is prime. Which numbers N should we try? Well, we can speed things up by restricting P to odd numbers, so we only have to try odd divisors N. When do we stop testing? When $N = P$? No, we can stop a lot sooner. In fact, we can stop once N reaches $\sqrt{P}$, since if there is a factor greater than $\sqrt{P}$ there must be a corresponding one less than $\sqrt{P}$, which we would have found. And where do we start? Well, since $N = 1$ will be a factor of any P, we should start at $N = 3$. The structure plan is as follows:

1. Input P
2. Initialize N to 3
3. Find remainder R when P is divided by N
4. While $R \neq 0$ and $N < \sqrt{P}$ repeat:
 Increase N by 2
 Find R when P is divided by N

5. If $R \neq 0$ then
 $\quad P$ is prime
 Else
 $\quad P$ is not prime
6. Stop.

Note that there may be no repeats—R might be zero the first time. Note also that there are *two* conditions under which the loop may stop. Consequently, an if is required after completion of the loop to determine which condition stopped it.

See if you can write the script. Then try it out on the following numbers: 4058879 (not prime), 193707721 (prime) and 2147483647 (prime). If such things interest you, the largest *known* prime number at the time of writing was $2^{6972593} - 1$ (discovered in June 1999). It has 2098960 digits and would occupy about 70 pages if it was printed in a newspaper. Obviously our algorithm cannot test such a large number, since it's unimaginably greater than the largest number which can be represented by MATLAB. Ways of testing such huge numbers for primality are described in D.E. Knuth, *The Art of Computer Programming. Volume 2: Seminumerical Algorithms* (Addison-Wesley, 1981). This particular whopper was found by the GIMPS (Great Internet Mersenne Prime Search). See http://www.utm.edu/research/primes/largest.html for more information on the largest known primes.

8.2.5 Projectile trajectory

In Chapter 3 we considered the flight of a projectile, given the usual equations of motion (assuming no air resistance). We would like to know now when and where it will hit the ground. Although this problem can be solved with a determinate loop (if you know enough applied mathematics), it is of interest also to see how to solve it with an indeterminate while loop. The idea is to calculate the trajectory repeatedly with increasing time, *while* the vertical displacement (y) remains positive. Here's the script:

```
dt = 0.1;
g = 9.8;
u = 60;
ang = input( 'Launch angle in degrees: ' );
ang = ang * pi / 180;            % convert to radians
x = 0; y = 0; t = 0;            % for starters
more(15)

while y >= 0
  disp( [t x y] );
  t = t + dt;
```

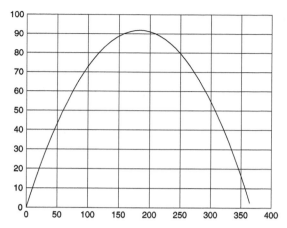

FIGURE 8.1
Projectile trajectory.

```
    y = u * sin(ang) * t - g * t^2 / 2;
    x = u * cos(ang) * t;
end
```

The command more(n) gives you n lines of output before pausing. This is called *paging*. To get another single line of output press **Enter**. To get the next page of n lines press the spacebar. To quit the script press **q**.

Try the script for some different launch angles. Can you find the launch angle which gives the maximum horizontal range (x)? What launch angle keeps it in the air for the longest time?

Note that when the loop finally ends, the value of y will be negative (check this by displaying y). However, the position of the disp statement ensures that only positive values of y are displayed. If for some reason you need to record the last value of t, say, before y becomes negative, you will need an if statement inside the while, e.g.,

```
if y >= 0
    tmax = t;
end
```

Change the script so that it displays the last time for which y was positive (tmax), *after* the while loop has ended.

Now suppose we want to plot the trajectory, as shown in Fig. 8.1. Note in particular how the trajectory stops above the x-axis. We need to use vectors now.

Here is the script:

```
dt = 0.1;
g = 9.8;
u = 60;
ang = input( 'Launch angle in degrees: ' );
ang = ang * pi / 180;            % convert to radians
xp = zeros(1); yp = zeros(1);    % initialize
y = 0; t = 0;
i = 1;                           % initial vector subscript

while y >= 0
  t = t + dt;
  i = i + 1;
  y = u * sin(ang) * t - g * t^2 / 2;
  if y >= 0
    xp(i) = u * cos(ang) * t;
    yp(i) = y;
  end
end

plot(xp, yp),grid
```

Note that the function `zeros` is used to initialize the vectors. This also clears any vector of the same name hanging around in the workspace from previous runs.

Note also the use of an `if` inside the `while` loop to ensure that only co-ordinates of points above the ground are added to the vectors `xp` and `yp`.

If you want the last point above the ground to be closer to the ground, try a smaller value of `dt`, e.g., 0.01.

8.2.6 `break` **and** `continue`

Any loop structure you are likely to encounter in scientific programming can be coded with either 'pure' `for` or `while` loops, as illustrated by the examples in this chapter. However, as a concession to intellectual laziness I feel obliged to mention in passing the `break` and `continue` statements.

If there are a number of different conditions to stop a `while` loop you may be tempted to use a `for` with the number of repetitions set to some accepted cut-off value (or even `Inf`) but enclosing `if` statements which `break` out of the `for` when the various conditions are met. Why is this not regarded as the best programming style? The reason is simply that when you read the code months later you will have to wade through the whole loop to find all the conditions

FIGURE 8.2
Menu window.

to end it, rather than see them all paraded at the start of the loop in the `while` clause.

If you are going to insist on using `break` you will have to look it up in `help` for yourself!

The `continue` statement is somewhat less virulent than `break` ...

8.2.7 Menus

Try the following program, which sets up a menu window, as in Fig. 8.2:

```
k = 0;

while k ~= 3
  k = menu( 'Click on your option', 'Do this', ...
          'Do that', 'Quit' );
  if k == 1
    disp( 'Do this ... press any key to continue ...' )
    pause
  elseif k == 2
    disp( 'Do that ... press any key to continue ...' )
    pause
  end
end
```

Note:

1. The `menu` function enables you to set up a menu of choices for a user.
2. `menu` takes only string arguments. The first one is the title of the menu. The second and subsequent strings are the choices available to the user.
3. The value returned by `menu` (`k` here) numbers the user's choices.
4. Since one has no idea how many choices the user will make, `menu` is properly enclosed in an indeterminate `while` loop. The loop continues to present the menu until the last option (in this example) is selected.

5. You can design much more sophisticated menu-driven applications with the MATLAB GUIDE (Graphical User Interface Development Environment).

SUMMARY

- A `for` statement should be used to program a determinate loop, where the number of repeats is known (in principle) *before* the loop is encountered. This situation is characterized by the general structure plan:

 > Repeat *N* times:
 > Block of statements to be repeated

 where *N* is known or computed *before* the loop is encountered for the first time, and is not changed by the block.
- A `while` statement should be used to program an indeterminate repeat structure, where the exact number of repeats is *not* known in advance. Another way of saying this is that these statements should be used whenever the truth value of the condition for repeating is changed in the body of the loop. This situation is characterized by the following structure plan:

 > While *condition* is true repeat:
 > statements to be repeated (reset truth value of *condition*).

 Note that *condition* is the condition to repeat.
- The statements in a `while` construct may sometimes never be executed.
- Loops may be nested to any depth.
- The `menu` statement inside a `while` loop may be used to present a user with a menu of choices.

EXERCISES

8.1 A person deposits $1000 in a bank. Interest is compounded monthly at the rate of 1 per cent per month. Write a program which will compute the monthly balance, but write it only *annually* for 10 years (use nested `for` loops, with the outer loop for 10 years, and the inner loop for 12 months). Note that after 10 years, the balance is $3300.39, whereas if interest had been compounded annually at the rate of 12 per cent per year the balance would only have been $3105.85.
See if you can vectorize your solution.

8.2 There are many formulae for computing π (the ratio of a circle's circumference to its diameter). The simplest is

$$\frac{\pi}{4} = 1 - 1/3 + 1/5 - 1/7 + 1/9 - \ldots \tag{8.4}$$

which comes from putting $x = 1$ in the series

$$\arctan x = x - \frac{x^3}{3} + \frac{x^5}{5} - \frac{x^7}{7} + \frac{x^9}{9} - \dots \tag{8.5}$$

(a) Write a program to compute π using Equation (8.4). Use as many terms in the series as your computer will reasonably allow (start modestly, with 100 terms, say, and re-run your program with more and more each time). You should find that the series converges very slowly, i.e. it takes a lot of terms to get fairly close to π.

(b) Rearranging the series speeds up the convergence:

$$\frac{\pi}{8} = \frac{1}{1 \times 3} + \frac{1}{5 \times 7} + \frac{1}{9 \times 11} \dots$$

Write a program to compute π using this series instead. You should find that you need fewer terms to reach the same level of accuracy that you got in (a).

(c) One of the fastest series for π is

$$\frac{\pi}{4} = 6 \arctan \frac{1}{8} + 2 \arctan \frac{1}{57} + \arctan \frac{1}{239}.$$

Use this formula to compute π. Don't use the MATLAB function `atan` to compute the arctangents, since that would be cheating. Rather use Equation (8.5).

(d) Can you vectorize any of your solutions (if you haven't already)?

8.3 The following method of computing π is due to Archimedes:

1. Let $A = 1$ and $N = 6$
2. Repeat 10 times, say:
 Replace N by $2N$
 Replace A by $[2 - \sqrt{(4 - A^2)}]^{1/2}$
 Let $L = NA/2$
 Let $U = L/\sqrt{1 - A^2/2}$
 Let $P = (U + L)/2$ (estimate of π)
 Let $E = (U - L)/2$ (estimate of error)
 Print N, P, E
3. Stop.

Write a program to implement the algorithm.

8.4 Write a program to compute a table of the function

$$f(x) = x \sin \left[\frac{\pi(1 + 20x)}{2} \right]$$

over the (closed) interval $[-1, 1]$ using increments in x of (a) 0.2, (b) 0.1 and (c) 0.01.

Use your tables to sketch graphs of $f(x)$ for the three cases (by hand), and observe that the tables for (a) and (b) give totally the wrong picture of $f(x)$.

Get your program to draw the graph of $f(x)$ for the three cases, super-imposed.

8.5 The transcendental number e (2.71828182845904 ...) can be shown to be the limit of

$$(1+x)^{1/x}$$

as x tends to zero (from above). Write a program which shows how this expression converges to e as x gets closer and closer to zero.

8.6 A square wave of period T may be defined by the function

$$f(t) = \begin{cases} 1 & (0 < t < T) \\ -1 & (-T < t < 0). \end{cases}$$

The Fourier series for $f(t)$ is given by

$$F(t) = \frac{4}{\pi} \sum_{k=0}^{\infty} \frac{1}{2k+1} \sin\left[\frac{(2k+1)\pi t}{T}\right].$$

It is of interest to know how many terms are needed for a good approximation to this infinite sum. Taking $T = 1$, write a program to compute and plot the sum to n terms of the series for t from -1.1 to 1.1 in steps of 0.01, say. Run the program for different values of n, e.g. 1, 3, 6, etc. Superimpose plots of $F(t)$ against t for a few values of n.

On each side of a discontinuity a Fourier series exhibits peculiar oscillatory behavior known as the Gibbs phenomenon. Fig. 8.3 shows this clearly for the above series with $n = 20$ (and increments in t of 0.01). The phenomenon is much sharper for $n = 200$ and t increments of 0.001.

8.7 If an amount of money A is invested for k years at a nominal annual interest rate r (expressed as a decimal fraction), the value V of the investment after k years is given by

$$V = A(1+r/n)^{nk}$$

where n is the number of compounding periods per year. Write a program to compute V as n gets larger and larger, i.e. as the compounding periods become more and more frequent, like monthly, daily, hourly, etc. Take $A = 1000$, $r = 4$ per cent and $k = 10$ years. You should observe that your output gradually approaches a limit. **Hint**: use a `for` loop which doubles n each time, starting with $n = 1$.

Also compute the value of the formula Ae^{rk} for the same values of A, r and k (use the MATLAB function `exp`), and compare this value with the values of V computed above. What do you conclude?

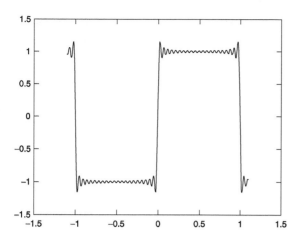

FIGURE 8.3
Fourier series: The Gibbs phenomenon.

8.8 Write a program to compute the sum of the series $1^2 + 2^2 + 3^2 \ldots$ such that the sum is as large as possible without exceeding 1000. The program should display how many terms are used in the sum.

8.9 One of the programs in Section 8.2 shows that an amount of $\$1000$ will double in eight years with an interest rate of 10 per cent. Using the same interest rate, run the program with initial balances of $\$500$, $\$2000$ and $\$10\,000$ (say) to see how long they all take to double. The results may surprise you.

8.10 Write a program to implement the structure plan of Exercise 3.2.

8.11 Use the Taylor series

$$\cos x = 1 - \frac{x^2}{2!} + \frac{x^4}{4!} - \frac{x^6}{6!} + \ldots$$

to write a program to compute $\cos x$ correct to four decimal places (x is in radians). See how many terms are needed to get 4-figure agreement with the MATLAB function cos. Don't make x too large; that could cause rounding error.

8.12 A student borrows $\$10\,000$ to buy a used car. Interest on her loan is compounded at the rate of 2 per cent per month while the outstanding balance of the loan is more than $\$5000$, and at 1 per cent per month otherwise. She pays back $\$300$ every month, except for the last month, when the repayment must be less than $\$300$. She pays at the end of the month, *after* the interest on the balance has been compounded. The first repayment is made one month after the loan is paid out. Write a program which displays a monthly statement of the balance (after the monthly payment has been made), the final payment, and the month of the final payment.

8.13 A projectile, the equations of motion of which are given in Chapter 4, is launched from the point O with an initial velocity of 60 m/s at an angle of 50° to the horizontal. Write a program which computes and displays the time in the air, and horizontal and vertical displacement from the point O every 0.5 seconds, as long as the projectile remains above a horizontal plane through O.

8.14 When a resistor (R), capacitor (C) and battery (V) are connected in series, a charge Q builds up on the capacitor according to the formula

$$Q(t) = CV(1 - e^{-t/RC})$$

if there is no charge on the capacitor at time $t = 0$. The problem is to monitor the charge on the capacitor every 0.1 seconds in order to detect when it reaches a level of 8 units of charge, given that $V = 9$, $R = 4$ and $C = 1$. Write a program which displays the time and charge every 0.1 seconds until the charge first exceeds 8 units (i.e. the last charge displayed must exceed 8). Once you have done this, rewrite the program to display the charge only while it is strictly less than 8 units.

8.15 Adapt your program for the prime number algorithm in Section 8.2 to find all the prime factors of a given number (even or odd).

MATLAB Graphics

CONTENTS

Basic 2-D
graphs 197
Labels 198
*Multiple plots on the
same axes* 199
*Line styles, markers
and color* 200
Axis limits 200
axes and axis? 201
*Multiple plots in a
figure:* subplot 202
figure, clf *and* cla 203
Graphical input 203
Logarithmic plots ... 203
Polar plots 204
*Plotting rapidly
changing
mathematical
functions:* fplot 205
The Property Editor 206

3-D plots 206
plot3 206
*Animated 3-D plots
with* comet3 207
Mesh surfaces 207
Contour plots 209
*Cropping a surface
with* NaNs 211
*Visualizing vector
fields* 211
*Visualization of
matrices* 212
*Rotation of 3-D
graphs* 213

Handle Graphics 214
Getting handles 215

THE OBJECTIVE OF THIS CHAPTER IS TO INTRODUCE YOU TO:

- MATLAB's high-level 2-D and 3-D plotting facilities
- Handle Graphics
- Editing plots
- Animation
- Saving and exporting graphs
- Color, lighting and the camera

A picture, it is said, is worth a thousand words. MATLAB has a powerful graphics system for presenting and visualizing data, which is reasonably easy to use. (Most of the figures in this book have been generated by MATLAB.)

This chapter introduces MATLAB's high-level 2-D and 3-D plotting facilities. Low-level features, such as Handle Graphics, are discussed later in this chapter.

It should be stressed that the treatment of graphics in this chapter is of necessity brief, and intended to give you a glimpse of the richness and power of MATLAB graphics. For the full treatment, you should consult help on the functions mentioned in this chapter, as well as the comprehensive list of graphics functions in Help document which you open up by clicking the '?' and typing graphics in the search tool.

9.1 BASIC 2-D GRAPHS

Graphs (in 2-D) are drawn with the plot statement. In its simplest form, it takes a single vector argument as in plot(y). In this case the elements of y are plotted against their indexes, e.g., plot(rand(1, 20)) plots 20 random numbers against the integers 1–20, joining successive points with straight lines, as in Fig. 9.1. If y is a matrix, its columns are plotted against element indexes.

Axes are automatically scaled and drawn to include the minimum and maximum data points.

Essential MATLAB for Engineers and Scientists. https://doi.org/10.1016/B978-0-08-102997-8.00015-4

Graphics object properties and how to change them 216
A vector of handles. 218
Graphics object creation functions .. 218
Parenting 219
Positioning figures . 219

Editing plots..... 220
Plot edit mode....... 220
Property Editor...... 221

Animation........ 222
Animation with Handle Graphics 223

Color etc. 225
Colormaps............ 225
Color of surface plots.................... 227
Truecolor 228

Lighting and camera 229

Saving, printing and exporting graphs 230
Saving and opening figure files 230
Printing a graph 230
Exporting a graph .. 230

Summary 231

Exercises 232

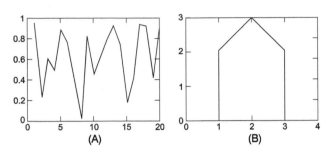

FIGURE 9.1
Examples of `plot`.

Probably the most common form of `plot` is `plot(x, y)` where `x` and `y` are vectors of the same length, e.g.,

```
x = 0:pi/40:4*pi;
plot(x, sin(x))
```

In this case, the co-ordinates of the ith point are x_i, y_i. This form of `plot` has been used widely in earlier chapters.

Straight-line graphs are drawn by giving the x and y co-ordinates of the end-points in two vectors. For example, to draw a line between the point with cartesian co-ordinates (0, 1) and (4, 3) use the statement

```
plot([0 4], [1 3])
```

i.e., `[0 4]` contains the x co-ordinates of the two points, and `[1 3]` contains their y co-ordinates.

MATLAB has a set of 'easy-to-use' plotting commands, all starting with the string 'ez'. The easy-to-use form of `plot` is `ezplot`, e.g.,

```
ezplot('tan(x)')
```

Exercises

1. Draw lines joining the following points: (0, 1), (4, 3), (2, 0) and (5, −2).
2. See if you can draw a 'house' similar to the one depicted in Fig. 9.1.

9.1.1 Labels

Graphs may be labeled with the following statements:

`gtext('text')`

writes a string (`'text'`) in the graph window. `gtext` puts a cross-hair in the graph window and waits for a mouse button or keyboard key to be

pressed. The cross-hair can be positioned with the mouse or the arrow keys. For example,

```
gtext( 'X marks the spot' )
```

Go on—try it!

Text may also be placed on a graph interactively with **Tools** -> **Edit Plot** from the figure window.

`grid`

adds/removes grid lines to/from the current graph. The grid state may be toggled.

`text(x, y, 'text')`

writes text in the graphics window at the point specified by x and y.

If x and y are vectors, the text is written at each point. If the text is an indexed list, successive points are labeled with corresponding rows of the text.

`title('text')`

writes the text as a title on top of the graph.

`xlabel('horizontal')`

labels the x-axis.

`ylabel('vertical')`

labels the y-axis.

9.1.2 Multiple plots on the same axes

There are at least three ways of drawing multiple plots on the same set of axes (which may however be rescaled if the new data falls outside the range of the previous data).

1. The easiest way is simply to use `hold` to keep the current plot on the axes. All subsequent plots are added to the axes until `hold` is released, either with `hold off`, or just `hold`, which toggles the hold state.
2. The second way is to use `plot` with multiple arguments, e.g.,

   ```
   plot(x1, y1, x2, y2, x3, y3, ... )
   ```

 plots the (vector) pairs $(x1, y1)$, $(x2, y2)$, etc. The advantage of this method is that the vector pairs may have different lengths. MATLAB automatically selects a different color for each pair.

 If you are plotting two graphs on the same axes you may find `plotyy` useful—it allows you to have independent y-axis labels on the left and the right, e.g.,

   ```
   plotyy(x,sin(x), x, 10*cos(x))
   ```

 (for x suitably defined).

3. The third way is to use the form

```
plot(x, y)
```

where x and y may both be matrices, or where one may be a vector and one a matrix.

If one of x or y is a matrix and the other is a vector, the rows or columns of the matrix are plotted against the vector, using a different color for each. Rows or columns of the matrix are selected depending on which have the same number of elements as the vector. If the matrix is square, columns are used.

If x and y are both matrices of the same size, the columns of x are plotted against the columns of y.

If x is not specified, as in plot(y), where y is a matrix, the columns of y are plotted against the row index.

9.1.3 Line styles, markers and color

Line styles, markers and colors may be selected for a graph with a string argument to plot, e.g.,

```
plot(x, y, '--')
```

joins the plotted points with dashed lines, whereas

```
plot(x, y, 'o')
```

draws circles at the data points with no lines joining them. You can specify all three properties, e.g.,

```
plot(x,sin(x), x, cos(x), 'om--')
```

plots $\sin(x)$ in the default style and color and $\cos(x)$ with circles joined with dashes in magenta. The available colors are denoted by the symbols c, m, y, k, r, g, b, w. You can have fun trying to figure out what they mean, or you can use help plot to see the full range of possible symbols.

9.1.4 Axis limits

Whenever you draw a graph with MATLAB it automatically scales the axis limits to fit the data. You can override this with

```
axis( [xmin, xmax, ymin, ymax] )
```

which sets the scaling on the *current* plot, i.e., draw the graph first, then reset the axis limits.

If you want to specify one of the minimum or maximum of a set of axis limits, but want MATLAB to autoscale the other, use `Inf` or `-Inf` for the autoscaled limit.

You can return to the default of automatic axis scaling with

```
axis auto
```

The statement

```
v = axis
```

returns the current axis scaling in the vector `v`.

Scaling is frozen at the current limits with

```
axis manual
```

so that if `hold` is turned on, subsequent plots will use the same limits.

If you draw a circle, e.g., with the statements

```
x = 0:pi/40:2*pi;
plot(sin(x), cos(x))
```

the circle probably won't appear round, especially if you resize the figure window. The command

```
axis equal
```

makes unit increments along the x- and y-axis the same physical length on the monitor, so that circles always appear round. The effect is undone with `axis normal`.

You can turn axis labeling and tick marks off with `axis off`, and back on again with `axis on`.

axes and axis?

You and I might be forgiven for thinking that the word 'axes' denotes the plural of 'axis', which it does indeed in common English usage. However, in MATLAB the word 'axes' refers to a particular graphics *object*, which includes not only the x-axis and y-axis and their tick marks and labels, but also everything drawn on those particular axes: the actual graphs and any text included in the figure. Axes objects are discussed in more detail later in this chapter.

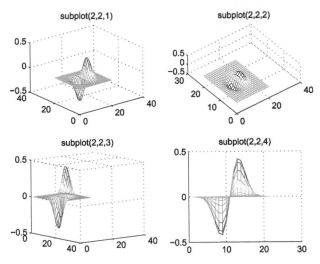

FIGURE 9.2
Four subplots: rotations of a 3-D surface.

9.1.5 Multiple plots in a figure: `subplot`

You can show a number of plots in the same figure window with the `subplot` function. It looks a little curious at first, but it's quite easy to get the hang of it. The statement

```
subplot(m, n, p)
```

divides the figure window into $m \times n$ small sets of axes, and selects the pth set for the current plot (numbered by row from the left of the top row). For example, the following statements produce the four plots shown in Fig. 9.2 (details of 3-D plotting are discussed in Section 9.2).

```
[x, y] = meshgrid(-3:0.3:3);
z = x .* exp(-x.^2 - y.^2);
subplot(2,2,1)
mesh(z),title('subplot(2,2,1)')
subplot(2,2,2)
mesh(z)
view(-37.5,70),title('subplot(2,2,2)')
subplot(2,2,3)
mesh(z)
view(37.5,-10),title('subplot(2,2,3)')
subplot(2,2,4)
mesh(z)
view(0,0),title('subplot(2,2,4)')
```

The command `subplot(1,1,1)` goes back to a single set of axes in the figure.

9.1.6 `figure`, `clf` **and** `cla`

`figure(h)`, where `h` is an integer, creates a new figure window, or makes figure `h` the current figure. Subsequent plots are drawn in the current figure. `h` is called the *handle* of the figure. Handle Graphics is discussed further in a later section of this chapter.

`clf` clears the current figure window. It also resets all properties associated with the axes, such as the hold state and the axis state.

`cla` deletes all plots and text from the current axes, i.e., leaves only the x- and y-axes and their associated information.

9.1.7 Graphical input

The command

```
[x, y] = ginput
```

allows you to select an unlimited number of points from the current graph using a mouse or arrow keys. A movable cross-hair appears on the graph. Clicking saves its co-ordinates in `x(i)` and `y(i)`. Pressing **Enter** terminates the input. An example is provided in the last chapter of this book which involves selecting points in a figure and fitting a curve to them.

The command

```
[x, y] = ginput(n)
```

works like `ginput` except that you must select exactly `n` points.

See `help` for further information.

9.1.8 Logarithmic plots

The command

```
semilogy(x, y)
```

plots y with a $\log_{10}$ scale and x with a linear scale. For example, the statements

```
x = 0:0.01:4;
semilogy(x, exp(x)), grid
```

produce the graph in Fig. 9.3. Equal increments along the y-axis represent multiples of powers of 10. So, starting from the bottom, the grid lines are drawn

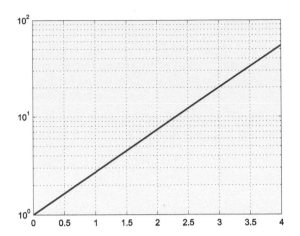

FIGURE 9.3
A logarithmic plot.

at 1, 2, 3, ..., 10, 20, 30, ..., 100, Incidentally, the graph of e^x on these axes is a straight line, because the equation $y = e^x$ transforms into a linear equation when you take logs of both sides.

See also `semilogx` and `loglog`.

Note that x and y may be vectors and/or matrices, just as in `plot`.

Exercise

Draw graphs of x^2, x^3, x^4 and e^{x^2} over the interval $0 \leq x \leq 4$, using `semilogy`.

9.1.9 Polar plots

The point (x, y) in cartesian co-ordinates is represented by the point (θ, r) in *polar* co-ordinates, where

$x = r \cos(\theta),$
$y = r \sin(\theta),$

and θ varies between 0 and 2π radians (360°).

The command

```
polar(theta, r)
```

generates a polar plot of the points with angles in `theta` and magnitudes in `r`.

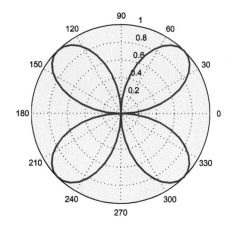

FIGURE 9.4
Polar plot of θ against $\sin(2\theta)$.

As an example, the statements

```
x = 0:pi/40:2*pi;
polar(x, sin(2*x)),grid
```

produce the plot shown in Fig. 9.4.

9.1.10 Plotting rapidly changing mathematical functions: fplot

In all the graphing examples so far, the x co-ordinates of the points plotted have been incremented uniformly, e.g., x = 0:0.01:4. If the function being plotted changes very rapidly in some places, this can be inefficient, and can even give a misleading graph.

For example, the statements

```
x = 0.01:0.001:0.1;
plot(x, sin(1./x))
```

produce the graph shown in Fig. 9.5A. But if the x increments are reduced to 0.0001, we get the graph in Fig. 9.5B instead. For $x < 0.04$, the two graphs look quite different.

MATLAB has a function called fplot which uses a more elegant approach. Whereas the above method evaluates $\sin(1/x)$ at equally spaced intervals, fplot evaluates it more frequently over regions where it changes more rapidly. Here's how to use it:

```
fplot('sin(1/x)', [0.01 0.1]) % no, 1./x not needed!
```

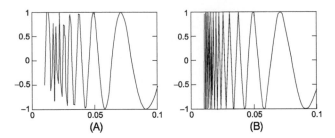

FIGURE 9.5

$y = \sin(1/x)$.

9.1.11 The Property Editor

The most general way of editing a graph is by using the Property Editor, e.g., **Edit** -> **Figure Properties** from the figure window. This topic is discussed briefly towards the end of this chapter.

9.2 3-D PLOTS

MATLAB has a variety of functions for displaying and visualizing data in 3-D, either as lines in 3-D, or as various types of surfaces. This section provides a brief overview.

9.2.1 plot3

The function `plot3` is the 3-D version of `plot`. The command

```
plot3(x, y, z)
```

draws a 2-D projection of a line in 3-D through the points whose co-ordinates are the elements of the vectors x, y and z. For example, the command

```
plot3(rand(1,10), rand(1,10), rand(1,10))
```

generates 10 random points in 3-D space, and joins them with lines, as shown in Fig. 9.6A.

As another example, the statements

```
t = 0:pi/50:10*pi;
plot3(exp(-0.02*t).*sin(t), exp(-0.02*t).*cos(t),t), ...
xlabel('x-axis'), ylabel('y-axis'), zlabel('z-axis')
```

produce the inwardly spiraling helix shown in Fig. 9.6B. Note the orientation of the x-, y- and z-axes, and in particular that the z-axis may be labeled with `zlabel`.

FIGURE 9.6
Examples of plot3.

9.2.2 Animated 3-D plots with comet3

The function comet3 is similar to plot3 except that it draws with a moving 'comet head'. Use comet3 to animate the helix in Fig. 9.6B. No prizes for guessing the name of the 2-D version.

9.2.3 Mesh surfaces

In Chapter 1 we saw how to draw the Mexican hat (Fig. 1.12):

```
[x y ] = meshgrid(-8 : 0.5 : 8);
r = sqrt(x.^2 + y.^2) + eps;
z = sin(r) ./ r;
```

This drawing is an example of a *mesh surface.*

To see how such surface is drawn, let's take a simpler example, say $z = x^2 - y^2$. The surface we are after is the one generated by the values of z as we move around the x-y plane. Let's restrict ourselves to part of the first quadrant of this plane, given by

$$0 \le x \le 5, \quad 0 \le y \le 5.$$

The first step is to set up the *grid* in the x-y plane over which the surface is to be plotted. You can use the MATLAB function meshgrid to do it, as follows:

```
[x y] = meshgrid(0:5);
```

This statement sets up two matrices, x and y. (Functions, such as meshgrid, which return more than one 'output argument', are discussed in detail in Chapter 10. However, you don't need to know the details in order to be able to use it here.)

The two matrices in this example are:

```
x =
      0    1    2    3    4    5
      0    1    2    3    4    5
      0    1    2    3    4    5
      0    1    2    3    4    5
      0    1    2    3    4    5
      0    1    2    3    4    5

y =
      0    0    0    0    0    0
      1    1    1    1    1    1
      2    2    2    2    2    2
      3    3    3    3    3    3
      4    4    4    4    4    4
      5    5    5    5    5    5
```

The effect of this is that the *columns* of the matrix x as it is displayed hold the x co-ordinates of the points in the grid, while the *rows* of the display of y hold the y co-ordinates. Recalling the way MATLAB array operations are defined, element by element, this means that the statement

```
z = x.^2 - y.^2
```

will correctly generate the surface points:

```
z =
       0     1     4     9    16    25
      -1     0     3     8    15    24
      -4    -3     0     5    12    21
      -9    -8    -5     0     7    16
     -16   -15   -12    -7     0     9
     -25   -24   -21   -16    -9     0
```

For example, at the grid point (5, 2), z has the value $5^2 - 2^2 = 21$. Incidentally, you don't need to worry about the exact relationship between grid co-ordinates and matrix subscripts; this is taken care of by meshgrid.

The statement mesh(z) then plots the surface (Fig. 9.7), with mesh lines connecting the points in the surface that lie above grid points.

Note that mesh(z) shows the row and column indices (subscripts) of the matrix z on the x and y axes. If you want to see proper values on the x and y axes use mesh(x,y,z). This applies to many of the other 3-D plotting functions.

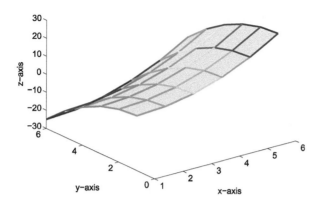

FIGURE 9.7
The surface $z = x^2 - y^2$.

The function mesh draws a surface as a 'wire frame'. An alternative visualization is provided by surf, which generates a faceted view of the surface (in color), i.e. the wire frame is covered with small tiles.

See help for variations on mesh and surf.

Exercises

1. Draw the surface shown in Fig. 9.7 with a finer mesh (of 0.25 units in each direction), using

   ```
   [x y] = meshgrid(0:0.25:5);
   ```

 (the number of mesh points in each direction is 21).
2. The initial heat distribution over a steel plate is given by the function

 $$u(x, y) = 80y^2 e^{-x^2 - 0.3y^2}.$$

 Plot the surface u over the grid defined by

 $$-2.1 \le x \le 2.1, \quad -6 \le y \le 6,$$

 where the grid width is 0.15 in both directions. You should get the plot shown in Fig. 9.8.

9.2.4 Contour plots

If you managed to draw the plot in Fig. 9.8, try the command

```
contour(u)
```

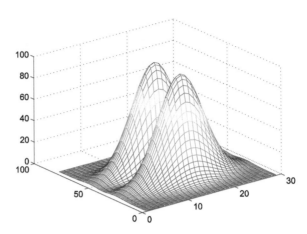

FIGURE 9.8
Heat distribution over a steel plate.

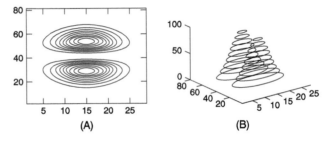

FIGURE 9.9
Contour plots.

You should get a *contour plot* of the heat distribution, as shown in Fig. 9.9A, i.e. the *isothermals* (lines of equal temperature). Here's the code:

```
[x y] = meshgrid(-2.1:0.15:2.1, -6:0.15:6); % x- y-grids different
u = 80 * y.^2 .* exp(-x.^2 - 0.3*y.^2);
contour(u)
```

The function `contour` can take a second input variable. It can be a scalar specifying how many contour levels to plot, or it can be a vector specifying the values at which to plot the contour levels.

You can get a 3-D contour plot with `contour3`, as shown in Fig. 9.9B.

Contour levels may be labeled with `clabel` (see `help`).

A 3-D contour plot may be drawn under a surface with `meshc` or `surfc`. For example, the statements

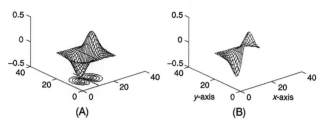

FIGURE 9.10
(A) meshc; (B) cropping a surface.

```
[x y] = meshgrid(-2:.2:2);
z = x .* exp(-x.^2 - y.^2);
meshc(z)
```

produce the graph in Fig. 9.10A.

9.2.5 Cropping a surface with NaNs

If a matrix for a surface plot contains NaNs, these elements are not plotted. This enables you to cut away (crop) parts of a surface. For example, the statements

```
[x y] = meshgrid(-2:.2:2, -2:.2:2);
z = x .* exp(-x.^2 - y.^2);
c = z;          % preserve the original surface
c(1:11,1:21) = NaN*c(1:11,1:21);
mesh(c), xlabel('x-axis'), ylabel('y-axis')
```

produce the graph in Fig. 9.10B.

9.2.6 Visualizing vector fields

The function quiver draws little arrows to indicate a gradient or other vector field. Although it produces a 2-D plot, it's often used in conjunction with contour, which is why it's described briefly here.

As an example, consider the scalar function of two variables $V = x^2 + y$. The *gradient* of V is defined as the *vector* field

$$\nabla V = \left(\frac{\partial V}{\partial x}, \frac{\partial V}{\partial y}\right)$$
$$= (2x, 1).$$

The following statements draw arrows indicating the direction of ∇V at points in the x-y plane (see Fig. 9.11):

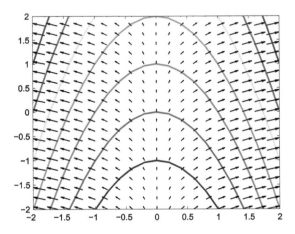

FIGURE 9.11
Gradients and level surfaces.

```
[x y] = meshgrid(-2:.2:2, -2:.2:2);
V = x.^2 + y;
dx = 2*x;
dy = dx;            % dy same size as dx
dy(:,:) = 1;        % now dy is same size as dx but all 1's
contour(x, y, V), hold on
quiver(x, y, dx, dy), hold off
```

The 'contour' lines indicate families of *level surfaces*; the gradient at any point is perpendicular to the level surface which passes through that point. The vectors x and y are needed in the call to contour to specify the axes for the contour plot.

An additional optional argument for quiver specifies the length of the arrows. See help.

If you can't (or don't want to) differentiate V, you can use the gradient function to estimate the derivative:

```
[dx dy] = gradient(V, 0.2, 0.2);
```

The values 0.2 are the increments in the x and y directions used in the approximation.

9.2.7 Visualization of matrices

The mesh function can also be used to 'visualize' a matrix. The following statements generate the plot in Fig. 9.12:

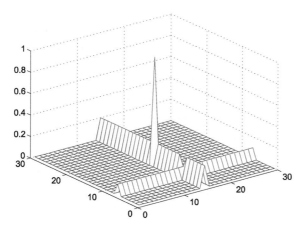

FIGURE 9.12
Visualization of a matrix.

```
a = zeros(30,30);
a(:,15) = 0.2*ones(30,1);
a(7,:) = 0.1*ones(1,30);
a(15,15) = 1;
mesh(a)
```

The matrix a is 30 × 30. The element in the middle—a(15,15)—is 1, all the elements in row 7 are 0.1, and all the remaining elements in column 15 are 0.2. mesh(a) then interprets the rows and columns of a as an *x-y* co-ordinate grid, with the values a(i,j) forming the mesh surface above the points (i, j).

The function spy is useful for visualizing sparse matrices.

9.2.8 Rotation of 3-D graphs

The view function enables you to specify the angle from which you view a 3-D graph. To see it in operation, run the following program, which rotates the visualized matrix in Fig. 9.12:

```
a = zeros(30,30);
a(:,15) = 0.2*ones(30,1);
a(7,:) = 0.1*ones(1,30);
a(15,15) = 1;
el = 30;
for az = -37.5:15:-37.5+360
  mesh(a), view(az, el)
  pause(0.5)
end
```

The function `view` takes two arguments. The first one, `az` in this example, is called the *azimuth* or polar angle in the *x-y* plane (in degrees). `az` rotates the *viewpoint* (you) about the *z*-axis—i.e. about the 'pinnacle' at (15,15) in Fig. 9.12—in a counter-clockwise direction. The default value of `az` is −37.5°. The program therefore rotates you in a counter-clockwise direction about the *z*-axis in 15° steps starting at the default position.

The second argument of `view` is the vertical elevation `el` (in degrees). This is the angle a line from the viewpoint makes with the *x-y* plane. A value of 90° for `el` means you are directly overhead. Positive values of the elevation mean you are above the *x-y* plane; negative values mean you are below it. The default value of `el` is 30°.

The command `pause(n)` suspends execution for n seconds.

You can rotate a 3-D figure interactively as follows. Click the Rotate 3-D button in the figure toolbar (first button from the right). Click on the axes and an outline of the figure appears to help you visualize the rotation. Drag the mouse in the direction you want to rotate. When you release the mouse button the rotated figure is redrawn.

Exercise

Rewrite the above program to change the elevation gradually, keeping the azimuth fixed at its default.

9.3 HANDLE GRAPHICS

The richness and power of MATLAB graphics is made possible by its *Handle Graphics objects*. There is a very helpful section devoted to Handle Graphics in the online documentation **MATLAB Help: Graphics**. What follows is of necessity a brief summary of its main features.

Handle Graphics objects are the basic elements used in MATLAB graphics. The objects are arranged in a parent-child inheritance structure as shown in Fig. 9.13. For example, Line and Text objects are children of Axes objects. This is probably the most common parent-child relationship. Axes objects define a region in a figure window and orient their children within this region. The actual plots in an Axes object are Line objects. The axis labels and any text annotations are Text objects. It is important to be aware of this parent-child hierarchy when you want to manipulate graphics objects using their handles.

Well, what exactly is the handle of a graphics object? Whenever MATLAB creates a graphics object it automatically creates a handle to that object. You can get the handle of the object using a function that explicitly returns the handle of a graphics object, or you can create the handle explicitly when you draw the graphics object. The handle itself has a floating point representation, but its

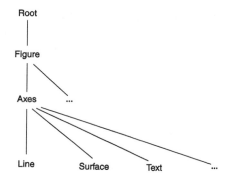

FIGURE 9.13
Parent-child relationships of Handle Graphics objects (from top to bottom).

actual value need not concern you. What is more important is to save its name and then to use the handle to change or manipulate your graphics object.

The root object is the only object whose handle is 0. There is only one root object, created by MATLAB at startup. All other objects are its descendants, as you can see in Fig. 9.13.

9.3.1 Getting handles

Here's how to get handles:

- The functions that draw graphics objects can also be used to return the handle of the object drawn, e.g.,

```
x = 0:pi/20:2*pi;
hsin = plot(x, sin(x))
hold on
hx = xlabel('x')
```

hsin is the handle of the Line object (the sine graph), and hx is the handle of the Text object (the *x* axis label) in the current Axes object.
The command

```
figure(h)
```

where h is an integer, creates a new figure, or makes figure h the current figure. h is the handle of the figure object.

- There are three functions that return the handle of particular graphics objects:

gcf gets the handle of the current figure, e.g.,

```
hf = gcf;
```

gca gets the handle of the current axes.

gco gets the handle of the current graphics object, which is the last graphics object created or clicked on. For example, draw the sine graph above and get its handle hsin. Click on the graph in the figure window. Then enter the command

```
ho = gco
```

in the Command Window. ho should be set to the handle of the sine graph (it should have the same numeric value as hsin).

9.3.2 Graphics object properties and how to change them

Once you have the handle of a graphics object you can use it to change the properties of that object. As an example, draw a sine graph and get its handle, as demonstrated above:

```
x = 0:pi/20:2*pi;
hsin = plot(x, sin(x))
```

Now suppose you want to make the graph much thicker. Enter the following command:

```
set(hsin, 'linewidth', 4);
```

You should get a nice fat sine curve!

linewidth is just one of the many properties of our graphics object. To see all the property names of an object and their current values use get(h) where h is the object's handle. In the case of our sine graph:

```
get(hsin)
 Color = [0 0 1]
 EraseMode = normal
 LineStyle = -
 LineWidth = [4]
 Marker = none
 MarkerSize = [6]
 MarkerEdgeColor = auto
 MarkerFaceColor = none
 XData = [ (1 by 41) double array]
 YData = [ (1 by 41) double array]
 ZData = []

 BeingDeleted = off
 ButtonDownFcn =
```

```
Children = []
Clipping = on
CreateFcn =
DeleteFcn =
BusyAction = queue
HandleVisibility = on
HitTest = on
Interruptible = on
Parent = [100.001]
Selected = off
SelectionHighlight = on
Tag =
Type = line
UIContextMenu = []
UserData = []
Visible = on
```

You can change any property value with the set function:

set(*handle*, '*PropertyName*', *PropertyValue*)

The command set(*handle*) lists all the possible property values (where appropriate).

You can get an object's handle and change its properties all in the same breath. For example,

```
set(gcf, 'visible', 'off')
```

makes the current figure invisible (without closing it—i.e., it's still 'there'). No prizes for guessing how to make it visible again ...

Property names are not case-sensitive, and you can abbreviate them to as few letters as make them unique. For example, you can abbreviate the type property to ty:

```
get(hsin,'ty')
ans =
line
```

(helpful if you don't know what type of objects you are dealing with!).

The different types of graphics objects don't all have the same set of properties, although there are some properties which are common to all graphics objects, such as children, parent, type, etc.

9.3.3 A vector of handles

If a graphics object has a number of children the `get` command used with the `children` property returns a vector of the children's handles. Sorting out the handles is then quite fun, and demonstrates why you need to be aware of the parent-child relationships!

As an example, plot a continuous sine graph and an exponentially decaying sine graph marked with o's in the same figure:

```
x = 0:pi/20:4*pi;
plot(x, sin(x))
hold on
plot(x, exp(-0.1*x).*sin(x), 'o')
hold off
```

Now enter the command

```
hkids = get(gca,'child')
```

You will see that a vector of handles with two elements is returned. The question is, which handle belongs to which plot? The answer is that the handles of children of the axes are returned in the *reverse order in which they are created*, i.e., `hkids(1)` is the handle of the exponentially decaying graph, while `hkids(2)` is the handle of the sine graph. So now let's change the markers on the decaying graph, and make the sine graph much bolder:

```
set(hkids(1), 'marker', '*')
set(hkids(2), 'linew', 4)
```

You should get the plots shown in Fig. 9.14.

If you are desperate and don't know the handles of any of your graphics objects you can use the `findobj` function to get the handle of an object with a property value that uniquely identifies it. In the original version of the plots in Fig. 9.14 the decaying plot can be identified by its `marker` property:

```
hdecay = findobj('marker', 'o' )
```

9.3.4 Graphics object creation functions

Each of the graphics objects shown in Fig. 9.13 (except the root object) has a corresponding creation function, named after the object it creates. See `help` for details.

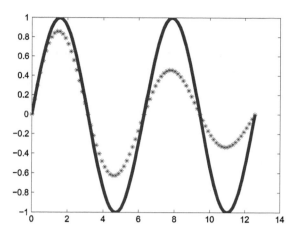

FIGURE 9.14
Result of manipulating a figure using the handles of axes children.

9.3.5 Parenting

By default, all graphics objects are created in the current figure. However, you can specify the parent of an object when you create it. For example,

```
axes('Parent', figure_handle, ... )
```

creates an axes in the figure with handle *figure_handle*. You can also move an object from one parent to another by redefining its `parent` property:

```
set(gca, 'Parent', figure_handle)
```

9.3.6 Positioning figures

At startup MATLAB determines the default position and size of the figure window, according to the size of your computer screen. You can change this by setting the `Position` property of the figure object.

Before you tamper with the figure's `Position` property you need to know the dimensions of your screen—this is one of the root object's properties. For example,

```
get(0, 'screensize')
ans =
      1     1    800    600
```

i.e., my screen is 800 by 600 pixels. The units of `Screensize` are pixels by defaults. You can (naturally) change the units by setting the root's `Units` property. For example,

```
set(0, 'units', 'normalized')
```

normalizes the width and height of the screen to 1. This is useful when you are writing an M-file which is to be run on different computer systems.

Having sorted out the size of your screen, you can then fiddle with the figure `Position` property, which is defined as a vector

```
[left bottom width height]
```

`left` and `bottom` define the position of the first addressable pixel in the lower left corner of the window, specified with respect to the lower left corner of the screen. `width` and `height` define the size of the interior of the window (excluding the window border).

You can normalize the figure's `Unit` property as well as the root's `Unit` property. You can then position figures absolutely, without reference to variations in screen size. For example, the following code normalizes units and creates two figures in the upper half of the screen:

```
set(0, 'units', 'normalized')
h1 = figure('units', 'normalized', 'visible', 'off')
h2 = figure('units', 'normalized', 'visible', 'off')
set(h1, 'position', [0.05 0.5 0.45 0.35], 'visible', 'on')
set(h2, 'position', [0.53 0.5 0.45 0.35], 'visible', 'on')
```

Note that the `Visible` property is first set to `off` to avoid the figures being drawn in the default position when they are created. They are only drawn when their positions have been redefined. Alternatively you can do everything when the figures are created:

```
h1 = figure('un', 'normalized', 'pos', [0.05 0.5 0.45 0.35])
```

9.4 EDITING PLOTS

There are a number of ways of editing plots.

9.4.1 Plot edit mode

To see how this works draw a graph, e.g., the friendly old sine. There are several ways to activate plot edit mode:

- Select **Tools** -> **Edit Plot** in the figure window.
- Click on the Edit Plot selection button in the figure window toolbar (arrow pointing roughly north-west).
- Run the `plotedit` command in the Command Window.

When a figure is in plot edit mode the toolbar selection button is highlighted. Once you are in plot edit mode select an object by clicking on it. Selection handles will appear on the selected object.

As an exercise, get the sine graph into plot edit mode and try the following:

- Select the graph (click on it). Selection handles should appear.
- Right click on the selected object (the graph). A context menu appears.
- Use the context menu to change the graph's line style and color.
- Use the **Insert** menu to insert a legend (although this makes more sense where you have multiple plots in the figure).
- Insert a text box inside the figure close to the graph as follows. Click on the Insert Text selection button in the toolbar (the capital A). The cursor changes shape to indicate that it is in text insertion mode. Move the insertion point until it touches the graph somewhere and click. A text box appears. Enter some text in it.
- Having labeled the graph you can change the format of the labels. Select the label and right-click. Change the font size and font style.
- Play around with the Insert Arrow and Insert Line selection buttons on the toolbar to see if you can insert lines and arrows on the graph.

To exit plot edit mode, click the selection button or uncheck the **Edit Plot** option on the **Tools** menu.

9.4.2 Property Editor

The Property Editor is more general than plot edit mode. It enables you to change object properties interactively, rather than with the set function. It is ideal for preparing presentation graphics.

There are numerous ways of starting the Property Editor (you may already have stumbled onto some):

- If plot edit mode is enabled you can:
 - Double-click on an object.
 - Right-click on an object and select **Properties** from the context menu.
- Select **Figure Properties**, **Axes Properties** or **Current Object Properties** from the figure **Edit** menu.
- Run the propedit command on the command line.

To experiment with the Property Editor it will be useful to have multiple plots in a figure:

```
x = 0:pi/20:2*pi;
hsin = plot(x,sin(x))
hold on
hcos = plot(x,cos(x))
hold off
```

Start the Property Editor and work through the following exercises:

- The navigation bar at the top of the Property Editor (labeled **Edit Properties for:**) identifies the object being edited. Click on the down arrow at the right of the navigation bar to see all the objects in the figure. You will notice that there are two line objects. Immediately we are faced with the problem of identifying the two line objects in our figure. The answer is to give each of them tags by setting their Tag properties.

 Go back to the figure and select the sine graph. Back in the Property Editor the navigation bar indicates you are editing one of the line objects. You will see three tabs below the navigation bar: **Data**, **Style** and **Info**. Click the **Info** tab, and enter a label in the **Tag** box, e.g., sine. Press **Enter**. The tag sine immediately appears next to the selected line object in the navigation bar.

 Give the cosine graph a tag as well (start by selecting the other line object).

- Select the sine graph. This time select the **Style** tab and go berserk changing its color, line style, line width and markers.

- Now select the axes object. Use the **Labels** tab to insert some axis labels. Use the **Scale** tab to change the *y* axis limits, for example.

 Note that if you were editing a 3-D plot you would be able to use the **Viewpoint** tab to change the viewing angle and to set various camera properties.

Have fun!

9.5 ANIMATION

There are three facilities for animation in MATLAB:

- The comet and comet3 functions can be used to draw comet plots, as we mentioned briefly in Chapter 7.

- The getframe function may be used to generate 'movie frames' from a sequence of graphs. The movie function can then be used to play back the movie a specified number of times.

 The MATLAB online documentation has the following script in **MATLAB Help: Graphics: Creating Specialized Plots: Animation**. It generates 16 frames from the Fast Fourier Transforms of complex matrices:

```
    for k = 1:16
        plot(fft(eye(k+16)))
        axis equal
        M(k) = getframe;
    end
```

Now play it back, say five times:

```
    movie(M, 5)
```

You can specify the speed of the playback, among other things. See `help`.
■ The most versatile (and satisfying) way of creating animations is by using the Handle Graphics facilities. Two examples follow.

9.5.1 Animation with Handle Graphics

For starters, run the following script, which should show the marker o tracing out a sine curve, leaving a trail behind it:

```
% animated sine graph
x = 0;
y = 0;
dx = pi/40;
p = plot(x, y, 'o', 'EraseMode', 'none'); % 'xor' shows only current point
                                           % ' none' shows all points
axis([0 20*pi -2 2])

for x = dx:dx:20*pi;
    x = x + dx;
    y = sin(x);
    set(p, 'XData', x, 'YData', y)
    drawnow
end
```

Note:

■ The statement

```
    p = plot(x, y, 'o', 'EraseMode', 'none');
```

achieves a number of things. It plots the first point of the graph. It saves the handle p of the plot for further reference. And it specifies that the Erasemode property is none, i.e., the object must not be erased when it is drawn again. To achieve complete animation, set this property to xor—try it now. Then the object is erased each time it is redrawn (in a slightly different position), creating the classic animation effect.

■ The statement

```
set(p, 'XData', x, 'YData', y)
```

sets the x and y data values of the object p to the new values generated in the for loop, and 'redraws' the object. However, it is not drawn on the screen immediately—it joins the 'event queue', where it waits until it is flushed out.

■ Finally, the drawnow function flushes the event queue and draws the object on the screen so that we can see the fruit of our labors.

As help drawnow will reveal, there are four events that flush the event queue:

- a return to the MATLAB prompt—this is how you have seen all the graphs you've drawn up to now;
- hitting a pause statement;
- executing a getframe command;
- executing a drawnow command.

For example, you could make the marker move in a more stately fashion by replacing drawnow with pause(0.05)—that's 0.05 seconds.

The next example is based on one in the **Animation** section of the MATLAB documentation. It involves *chaotic* motion described by a system of three non-linear differential equations having a 'strange attractor' (known as the Lorenz strange attractor). The system can be written as

$$\frac{dy}{dt} = \mathbf{A}y,$$

where $y(t)$ is a vector with three components, and $\mathbf{A}$ is a matrix depending on y:

$$\mathbf{A}(y) = \begin{bmatrix} -8/3 & 0 & y(2) \\ 0 & -10 & 10 \\ -y(2) & 28 & -1 \end{bmatrix}.$$

The following script solves the system approximately using Euler's method (a more accurate numerical solution is discussed in Chapter 17), and shows the solution orbiting about two different attractors without settling into a steady orbit about either. Fig. 9.15 shows the situation after a few thousand points have been plotted.

```
A = [ -8/3 0 0; 0 -10 10; 0 28 -1 ];
y = [35 -10 -7]';
h = 0.01;
p = plot3(y(1), y(2), y(3), 'o', ...
          'erasemode', 'none', 'markersize', 2);
```

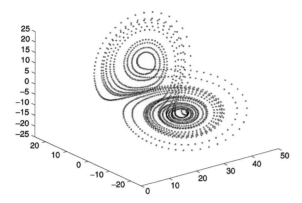

FIGURE 9.15
The Lorenz strange attractor.

```
axis([0 50 -25 25 -25 25])
hold on
i = 1;

while 1
    A(1,3) = y(2);
    A(3,1) = -y(2);
    ydot = A*y;
    y = y + h*ydot;
    % Change color occasionally
    if rem(i,500) == 0
        set(p, 'color', [rand, rand, rand])
    end
    % Change co-ordinates
    set(p, 'XData', y(1), 'YData', y(2), 'ZData', y(3))
    drawnow
    i=i+1;
end
```

If all the points are all plotted in the same color you eventually wouldn't be able to see the new points generated: a large area of the screen would be filled with the drawing color. The color is therefore set randomly after every 500 points are plotted.

9.6 COLOR ETC.

9.6.1 Colormaps

MATLAB graphics can generate a rich variety of colors. You may be wondering how.

The following script shows a view of the Earth from space:

```
load earth
image(X); colormap(map)
axis image
```

(axis image is the same as axis equal except that the plot box fits tightly around the data). For a variation use hot as the argument of colormap instead of map.

The matrix X loaded from earth is 257-by-250. Each of the elements of X is an integer in the range 1–64. For example, here is a 3-by-3 submatrix of X (somewhere in north-east Africa):

```
X(39:41,100:102)
ans =
    14    15    14
    10    16    10
    10    10    10
```

The colormap function generates by default a 64-by-3 matrix with elements in the range 0–1. The values in the three columns represent the intensities of the red, green and blue (RGB) video components respectively. Each row of this matrix therefore defines a particular color by specifying its RGB components. The image function maps each element of its argument to a row in the colormap to find the color of that element. For example, X(40,101) has the value 16. Row 16 of the colormap has the three values

```
    0.6784    0.3216    0.1922
```

(reddish) as you can easily verify with the statements

```
cm = colormap(map);
cm(16,:)
```

(The map colormap is also loaded from earth.) These values of RGB specify the color of the pixel 40 from the top and 101 from the left in the figure. Incidentally, you can use the statement

```
[xp yp] = ginput
```

to get the co-ordinates of a point in the figure (cross-hairs appears in the figure; click on the point whose co-ordinates you want). Clicking on the point in north-east Africa results in

```
xp =
    101.8289  % from the left (column of X)
yp =
     40.7032  % from the top (row of X)
```

Note that xp and yp correspond to the *columns* and *rows* of X respectively (imagine the matrix X superimposed on the figure).

There are a number of functions provided by MATLAB which generate colormaps, e.g., jet (the default), bone, flag, prism. See help graph3d for a complete list.

You can sample the various colormaps quite nicely with the following statement, which shows 64 vertical strips, each in a different color:

```
image(1:64),colormap(prism)
```

Or why not generate some random colors?

```
randmap(:,1) = rand(64,1);
randmap(:,2) = rand(64,1);
randmap(:,3) = rand(64,1);
image(1:64);colormap(randmap)
```

The function colorbar displays the current colormap vertically or horizontally in the figure with your graph, indicating how the 64 colors are mapped. Try it with the image of the Earth.

Note that 64 is the default length of a colormap. The functions that generate colormaps have an optional parameter specifying the length of the colormap.

9.6.2 Color of surface plots

When you draw a surface plot with a single matrix argument, e.g., surf(z), the argument z specifies both the height of the surface and the color. As an example use the function peaks which generates a surface with a couple of peaks and valleys:

```
z = peaks;
surf(z), colormap(jet), colorbar
```

The colorbar indicates that the minimum element of z (somewhat less than -6) is mapped to row 1 of the colormap ($R = 0$, $G = 0$, $B = 0.5625$), whereas the maximum element (about 8) is mapped to row 64 ($R = 0.5625$, $G = 0$, $B = 0$).

You can specify the color with a second argument the same size as the first:

```
z = peaks(16);   % generates a 16-by-16 mesh
c = rand(16);
surf(z, c), colormap(prism)
```

Here the surface is tiled with a random pattern of 16-by-16 colors from the prism colormap.

In this form of surf each element of c is used to determine the color of the point in the corresponding element of z. By default MATLAB uses a process called *scaled mapping* to map from an element of z (or c) to the color in the colormap. The details of the scaling are determined by the caxis command. For further details see help caxis or the section **Coloring Mesh and Surface Plots** in **MATLAB Help: 3-D Visualization: Creating 3-D Graphs**.

You can exploit the facility to specify color in order to emphasize properties of the surface. The following example is given in the MATLAB documentation:

```
z = peaks(40);
c = del2(z);
surf(z, c)
colormap hot
```

The function del2 computes the discrete Laplacian of a surface—the Laplacian is related to the curvature of the surface. Creating a color array from the Laplacian means that regions with similar curvature will be drawn in the same color. Compare the surface obtained this way with that produced by the statement

```
surf(P), colormap(hot)
```

In the second case regions with similar heights about the *x-y* plane have the same color.

Alternative forms of surf (and related surface functions) are:

```
surf(x, y, z)        % color determined by z
surf(x, y, z, c)     % color determined by c
```

9.6.3 Truecolor

The technique of coloring by means of a colormap is called *indexed* coloring—a surface is colored by assigning each data point an index (row) in the color map. The Truecolor technique colors a surface using explicitly specified RGB triplets. Here is another example from the MATLAB documentation (it also demonstrates the use of a multi-dimensional array):

```
z = peaks(25);
c(:,:,1) = rand(25);
c(:,:,2) = rand(25);
c(:,:,3) = rand(25);
surf(z, c)
```

The three 'pages' of c (indicated by its third subscript) specify the values of RGB respectively to be used to color the point in z whose subscripts are the same as the first two subscripts in c. For example, the RGB values for the color of the point $z(5,13)$ are given by $c(5,13,1)$, $c(5,13,2)$ and $c(5,13,3)$ respectively.

9.7 LIGHTING AND CAMERA

MATLAB uses lighting to add realism to graphics, e.g., illuminating a surface by shining light on it from a certain angle. Here are two examples from the MATLAB documentation. Have a look at them.

```
z = peaks(25);
c(:,:,1) = rand(25);
c(:,:,2) = rand(25);
c(:,:,3) = rand(25);
surf(z,c,'facecolor','interp','facelighting','phong',...
        'edgecolor','none')
camlight right
```

The possibilities for the `facelighting` property of the surface object are `none`, `flat` (uniform color across each facet), `gouraud` or `phong`. The last two are the names of lighting algorithms. Phong lighting generally produces better results, but takes longer to render than Gouraud lighting. Remember that you can see all a surface object's properties by creating the object with a handle and using `get` on the object's handle.

This one's quite stunning:

```
[x y ] = meshgrid(-8 : 0.5 : 8);
r = sqrt(x.^2 + y.^2) + eps;
z = sin(r) ./ r;
surf(x,y,z,'facecolor','interp','edgecolor','none', ...
        'facelighting','phong')
colormap jet
daspect([5 5 1])
axis tight
view(-50, 30)
camlight left
```

For more information on lighting and the camera see the sections **Lighting as a Visualization Tool** and **Defining the View** in MATLAB Help: 3-D Visualization.

9.8 SAVING, PRINTING AND EXPORTING GRAPHS

9.8.1 Saving and opening figure files

You can save a figure generated during a MATLAB session so that you can open it in a subsequent session. Such a file has the .fig extension.

- Select **Save** from the figure window **File** menu.
- Make sure the **Save as type** is .fig.

To open a figure file select **Open** from the **File** menu.

9.8.2 Printing a graph

You can print everything inside a figure window frame, including axis labels and annotations:

- To print a figure select **Print** from the figure window **File** menu.
- If you have a black and white printer, colored lines and text are 'dithered to gray' which may not print very clearly in some cases. In that case select **Page Setup** from the figure **File** menu. Select the **Lines and Text** tab and click on the **Black and white** option for **Convert solid colored lines to:**.

9.8.3 Exporting a graph

A figure may be exported in one of a number of graphics formats if you want to import it into another application, such as a text processor. You can also export it to the Windows clipboard and paste it from there into an application.

To export to the clipboard:

- Select **Copy Figure** from the figure window's **Edit** menu (this action copies to the clipboard).
- Before you copy to the clipboard you may need to adjust the figure's settings. You can do this by selecting **Preferences** from the figure's **File** menu. This action opens the **Preferences** panel, from which you can select **Figure Copy Template Preferences** and **Copy Options Preferences** to adjust the figure's settings.
 You may also have to adjust further settings with **Page Setup** from the figure's **File** menu.

To export a figure to a file in a specific graphics format:

- Select **Export** from the figure's **File** menu. This action invokes the **Export** dialogue box.
- Select a graphics format from the **Save as type** list, e.g., EMF (enhanced metafiles), JPEG, etc. You may need to experiment to find the best format for the target application you have in mind.

 For example, to insert a figure into a Word document, I find it much easier first to save it in EMF or JPEG format and then to insert the graphics file into the Word document, rather than to go the clipboard route (there are more settings to adjust that way).

For further details consult the section **Basic Printing and Exporting** in **MATLAB Help: Graphics**.

SUMMARY

- 2-D plots are drawn with the `plot` statement.
- There is a set of easy-to-use plotters called `ez*`.
- Graphs may be labeled with `grid`, `text`, `title`, `xlabel`, `ylabel`, etc.
- Multiple plots may be drawn on the same axes in a number of ways.
- Line styles, markers and color may be varied.
- Axis limits may be set explicitly.
- In MATLAB the word 'axes' refers to the graphics object in which the x and y axis and their labels, plots and text annotations are drawn.
- A number of axes may be drawn in the same figure with `subplot`.
- Co-ordinates of points in a figure may be selected with `ginput`.
- `semilogx`, `semilogy` and `loglog` are used to plot on $\log_{10}$ scales.
- The `polar` command plots in polar co-ordinates.
- `fplot` provides a handy way of plotting mathematical functions.
- `plot3` draws lines in 3-D.
- `comet3` animates a 3-D plot.
- A 3-D surface may be plotted with `mesh`.
- A 3-D plot may be rotated with `view` or with the **Rotate 3-D** tool in the figure window.
- `mesh` may also be used to visualize a matrix.
- `contour` and `contour3` draws contour levels in 2-D and 3-D respectively.
- 3-D surfaces may be cropped.
- For a complete list of graphics functions consult the online Help in **MATLAB Function Reference: Functions by Category: Graphics**.
- MATLAB graphics objects are arranged in a parent-child hierarchy.
- A handle may be attached to a graphics object at creation; the handle may be used to manipulate the graphics object.

- If h is the handle of a graphics object, get(h) returns all the current values of the object's properties. set(h) shows you all the possible values of the properties.
- The functions gcf, gca and gco return the handles of various graphics objects.
- Use set to change the properties of a graphics object.
- A graph may be edited to a limited extent in plot edit mode, selected from the figure window (**Tools** -> **Edit Plot**). More general editing can be done with the Property Editor (**Edit** -> **Figure Properties** from the figure window).
- The most versatile way to create animations is by using the Handle Graphics facilities. Other techniques involve comet plots and movies.
- Indexed coloring may be done with colormaps.
- Graphs saved to .fig files may be opened in subsequent MATLAB sessions.
- Graphs may be exported to the Windows clipboard, or to files in a variety of graphics formats.

EXERCISES

9.1 Draw a graph of the population of the USA from 1790 to 2000, using the (logistic) model

$$P(t) = \frac{197\,273\,000}{1 + e^{-0.03134(t - 1913.25)}}$$

where t is the date in years.

Actual data (in 1000s) for every decade from 1790 to 1950 are as follows: 3929, 5308, 7240, 9638, 12\,866, 17\,069, 23\,192, 31\,443, 38\,558, 50\,156, 62\,948, 75\,995, 91\,972, 1\,05\,711, 1\,22\,775, 1\,31\,669, 1\,50\,697. Superimpose this data on the graph of $P(t)$. Plot the data as discrete circles (i.e. do not join them with lines) as shown in Fig. 9.16.

9.2 The Spiral of Archimedes (Fig. 9.17) may be represented in polar coordinates by the equation

$$r = a\theta,$$

where a is some constant. (The shells of a class of animals called nummulites grow in this way.) Write some command-line statements to draw the spiral for some values of a.

9.3 Another type of spiral is the *logarithmic* spiral (Fig. 9.17), which describes the growth of shells of animals like the periwinkle and the

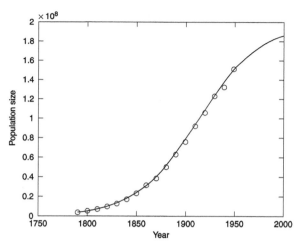

FIGURE 9.16
USA population: model and census data (o).

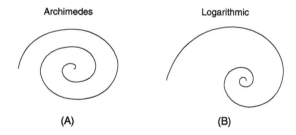

FIGURE 9.17
Spirals.

nautilus. Its equation in polar co-ordinates is

$$r = aq^\theta,$$

where $a > 0$, $q > 1$. Draw this spiral.

9.4 The arrangement of seeds in a sunflower head (and other flowers, like daisies) follows a fixed mathematical pattern. The nth seed is at position

$$r = \sqrt{n},$$

with angular co-ordinate $\pi dn/180$ radians, where d is the constant angle of divergence (in degrees) between any two successive seeds, i.e. between the nth and $(n+1)$th seeds. A perfect sunflower head (Fig. 9.18) is generated by $d = 137.51°$. Write a program to plot the seeds; use a circle (o) for each seed. A remarkable feature of this model is that the angle d must be exact to get proper sunflowers. Experiment with some different values, e.g. 137.45° (spokes, from fairly far out), 137.65° (spokes all the way), 137.92° (Catherine wheels).

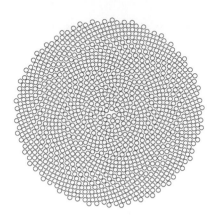

FIGURE 9.18
A perfect sunflower?

9.5 The equation of an ellipse in polar co-ordinates is given by

$$r = a(1 - e^2)/(1 - e\cos\theta),$$

where a is the semi-major axis and e is the eccentricity, if one focus is at the origin, and the semi-major axis lies on the x-axis.

Halley's Comet, which visited us in 1985/6, moves in an elliptical orbit about the Sun (at one focus) with a semi-major axis of 17.9 A.U. (A.U. stands for Astronomical Unit, which is the mean distance of the Earth from the Sun: 149.6 million km.) The eccentricity of the orbit is 0.967276. Write a program which draws the orbit of Halley's Comet and the Earth (assume the Earth is circular).

9.6 A very interesting iterative relationship that has been studied a lot recently is defined by

$$y_{k+1} = r y_k (1 - y_k)$$

(this is a discrete form of the well-known *logistic model*). Given y_0 and r, successive y_k's may be computed very easily, e.g. if $y_0 = 0.2$ and $r = 1$, then $y_1 = 0.16$, $y_2 = 0.1334$, and so on.

This formula is often used to model population growth in cases where the growth is not unlimited, but is restricted by shortage of food, living area, etc.

y_k exhibits fascinating behavior, known as *mathematical chaos*, for values of r between 3 and 4 (independent of y_0). Write a program which plots y_k against k (as individual points).

Values of r that give particularly interesting graphs are 3.3, 3.5, 3.5668, 3.575, 3.5766, 3.738, 3.8287, and many more that can be found by patient exploration.

9.7 A rather beautiful *fractal* picture can be drawn by plotting the points (x_k, y_k) generated by the following difference equations

$$x_{k+1} = y_k(1 + \sin 0.7x_k) - 1.2\sqrt{|x_k|},$$
$$y_{k+1} = 0.21 - x_k,$$

starting with $x_0 = y_0 = 0$. Write a program to draw the picture (plot individual points; do not join them).

Vectors as Arrays and Other Data Structures

THE OBJECTIVES OF THIS CHAPTER ARE:

- To enable you to solve problems which involve working carefully with array subscripts.
- To introduce you to:
 - Structures.
 - Cells and cell arrays.

In MATLAB an array is just another name for a vector. So why have a large part of a chapter on arrays, when we have already been using vectors for most of the book? It is helpful to talk about arrays (as opposed to vectors) when we are concerned with handling individual elements by means of their subscripts, rather than with the vector as a whole. Thus, in the first three sections of this chapter we look at a number of problems which are best solved by treating vectors as arrays, usually with the help of `for` loops.

In the last three sections of this chapter we deal with other, more advanced, data structures.

10.1 UPDATE PROCESSES

In Chapter 8 (**Update processes**) we considered the problem of calculating the temperature of orange juice (OJ) in a can as it cools in a fridge. This is an example of an *update process*, where the main variable is repeatedly updated over a period of time. We now examine how to solve this problem more generally.

The OJ is initially at temperature 25°C when it is placed in the fridge, where the ambient temperature F is 10°. The standard way to solve such an update process is to break the time period up into a number of small steps, each of length dt. If T_i is the temperature at the beginning of step i, we can get T_{i+1} from T_i as follows:

$$T_{i+1} = T_i - K\,dt\,(T_i - F), \tag{10.1}$$

where K is a physical constant, and units are chosen so that time is in minutes.

CONTENTS

Update processes 237
Unit time steps 238
Non-unit time steps 240
Using a function 242
Exact solution 243

Frequencies, bar charts and histograms 244
A random walk 244
Histograms 245

Sorting 246
Bubble Sort 246
MATLAB's `sort` 248

Structures 249

Cell arrays 251
Assigning data to cell arrays 251
Accessing data in cell arrays 253
Using cell arrays 253
Displaying and visualizing cell arrays 254

Classes and objects 254

Summary 255

Essential MATLAB for Engineers and Scientists. https://doi.org/10.1016/B978-0-08-102997-8.00016-6

10.1.1 Unit time steps

We first solve the problem using a unit time step, i.e., $dt = 1$. The simplest way is to use scalars for the time and the temperature, as we saw in Chapter 8 (although we didn't use a unit time step there):

```
K = 0.05;
F = 10;
T = 25;                       % initial temperature of OJ

for time = 1:100              % time in minutes
  T = T - K * (T - F);        % dt = 1
  if rem(time, 5) == 0
    disp( [time T] )
  end
end;
```

Note the use of rem to display the results every 5 minutes: when time is an integer multiple of 5 its remainder when divided by 5 will be zero.

While this is undoubtedly the easiest way of writing the script, we cannot easily plot the graph of temperature against time. In order to do that, time and T must be vectors. The index of the for loop must be used as the subscript of each element of T. Here's the script (update1.m):

```
K = 0.05;
F = 10;
time = 0:100;                 % initialize vector time
T = zeros(1,101);             % pre-allocate vector T
T(1) = 25;                    % initial temperature of OJ

for i = 1:100                           % time in minutes
  T(i+1) = T(i) - K * (T(i) - F);  % construct T
end;

disp([ time(1:5:101)' T(1:5:101)' ]);  % display results
plot(time, T), grid                     % every 5 mins
```

See Fig. 10.1 for typical graphs.

Note:

- The statement time = 0:100 sets up a (row) vector for time where time(1) has the value 0 minutes, and time(101) has the value 100 minutes. This is necessary because the first subscript of a MATLAB vector must be 1.

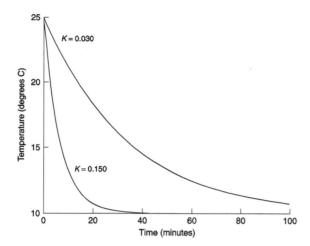

FIGURE 10.1
Cooling curves.

- The statement `T = zeros(1,101)` sets up a corresponding (row) vector for the temperature, with every element initialized to zero (again there must be 101 elements, because the first element is for the temperature at time zero). This process is called *pre-allocation*. It serves two important purposes.

 (a) Firstly, it clears a vector of the same name left over from a previous run. This could cause a conflict when attempting to display or plot T against `time`, if the vectors have different sizes. To see this, run `update1` as it stands. It should work perfectly. Now suppose you decide to do the calculations over a shorter time period, say 50 minutes. Remove the `zeros` statement, and make the following additional two changes and re-run the script (but *don't* clear the workspace):

  ```
  time = 0:50;           % initialize vector time
  ...

  for i = 1:50           % time in minutes
  ```

 This time you get an error message:

  ```
  ??? Error using ==> plot
  Vectors must be the same lengths.
  ```

 `whos` will reveal that `time` is correctly 51-by-1, but T is still 101-by-1. The `plot` command naturally requires these vectors to have the same lengths.

 The problem is that, while the operation `0:50` redefines `time` correctly, the `for` loop does not have the same effect on T. Since T is being updated element by element, the unused elements from 51

to 101 from the previous run are left untouched in the workspace. Pre-allocation of the correct number of elements to T with zeros avoids this problem.

(b) Secondly, although the script will work without the zeros statement, as we have seen, it will be much *slower*, since then T has to be *redimensioned* during each repeat of the for loop, in order to make space for a new element each time.

It's extremely instructive to do an experiment using a vector size of 10000 elements, say, i.e.,

```
time = 0:9999;          % initialize vector time
T = zeros(1,10000);     % pre-allocate vector T
...
for i = 1:10000
...
```

(and comment out the disp and plot statements, since these will obscure the issue). My Pentium II takes 0.99 seconds to run the script with pre-allocation of T, but 13.19 seconds to run without pre-allocation—more than 10 times longer. This could be a critical consideration in a script which does a lot of such element-by-element processing.

- The first element of T is set to the initial temperature of the OJ. This is the temperature at time zero.
- The for loop computes the values for T(2), ..., T(101).
 This arrangement ensures that temperature T(i) corresponds to time(i).
- The colon operator is used to display the results at 5-minute intervals.

10.1.2 Non-unit time steps

It's not always appropriate and/or accurate enough to take $dt = 1$ in Equation (10.1). There is a standard way of generating the solution vector in MATLAB, given (almost) any value of dt. We introduce a more general notation to do this.

Call the initial time a, and the final time b. If we want time steps of length dt, the number m of such steps will be

$$m = (b - a)/dt.$$

The time at the *end* of step i will therefore be $a + i\,dt$.

The following script, update2.m, implements this scheme. It has some additional features. It prompts you for a value of dt, and checks that this value gives an integer number of steps m. It also asks you for the output interval opint (the intervals in minutes at which the results are displayed in table form) and

checks that this interval is an integer multiple of *dt*. Try the script with same sample values, e.g., *dt* = 0.4 and opint = 4.

```
K = 0.05;
F = 10;
a = 0;                    % initial time
b = 100;                  % final time
load train
dt = input( 'dt: ' );
opint = input( 'output interval (minutes): ' );
if opint/dt ~= fix(opint/dt)
  sound(y, Fs)
  disp( 'output interval is not a multiple of dt!' )
  break
end;

m = (b - a) / dt;      % m steps of length dt
if fix(m) ~= m         % make sure m is integer
  sound(y, Fs)
  disp( 'm is not an integer - try again!' );
  break
end;

T = zeros(1,m+1);      % pre-allocate (m+1) elements
time = a:dt:b;
T(1) = 25;             % initial temperature

for i = 1:m
  T(i+1) = T(i) - K * dt * (T(i) - F);
end;

disp( [time(1:opint/dt:m+1)' T(1:opint/dt:m+1)'] )
plot(time, T),grid
```

Note:

- The vectors T and time must each have $m + 1$ elements, because there are *m* time steps, and we need an extra element for the initial value of each vector.
- The expression opint/dt gives the index increment for displaying the results, e.g., *dt* = 0.1 and opint = 0.5 displays every (0.5/0.1)th element, i.e., every 5th element.

10.1.3 Using a function

The really cool way of solving this problem is to write a function to do it. This makes it a lot easier to generate a table of results for different values of *dt*, say, using the function from the command line. Here is update2.m rewritten as a function cooler.m:

```
function [time, T, m] = cooler( a, b, K, F, dt, T0 )

m = (b - a) / dt;      % m steps of length dt
if fix(m) ~= m         % make sure m is integer
  disp( 'm is not an integer - try again!' );
  break
end;

T = zeros(1,m+1);      % pre-allocate
time = a:dt:b;
T(1) = T0;             % initial temperature

for i = 1:m
  T(i+1) = T(i) - K * dt * (T(i) - F);
end;
```

Suppose you want to display a table of temperatures against time at 5 minute intervals, using *dt* = 1 and *dt* = 0.1. Here is how to do it (in the Command Window):

```
dt = 1;
[t T m] = cooler(0, 100, 0.05, 10, dt, 25);
table(:,1) = t(1:5/dt:m+1)';
table(:,2) = T(1:5/dt:m+1)';
dt = 0.1;
[t T m] = cooler(0, 100, 0.05, 10, dt, 25);
table(:,3) = T(1:5/dt:m+1)';
format bank
disp(table)
```

Output:

```
      0         25.00         25.00
   5.00         21.61         21.67
  10.00         18.98         19.09
    ...
 100.00         10.09         10.10
```

Note:

- The advantage of using a function which generates a vector output variable is that even if you forget to pre-allocate the vector inside the function (with `zeros`) MATLAB automatically clears any previous versions of the output vector before returning from the function.
- The variable `table` is a two-dimensional array (or matrix). Recall that the colon operator may be used to indicate all elements of a matrix in a row or column. So `table(:,1)` means the elements in every row and column 1, i.e., the entire first column. The vectors `t` and `T` are row vectors, so they must be transposed before being inserted into columns 1 and 2 of `table`.
 The third column of `table` is inserted in a similar way.
- The results in the third column (for $dt = 0.1$) will be more accurate.

10.1.4 Exact solution

This cooling problem has an exact mathematical solution. The temperature $T(t)$ at time t is given by the formula

$$T(t) = F + (T_0 - F)e^{-Kt}, \tag{10.2}$$

where T_0 is the initial temperature. You can insert values for this exact solution into a fourth column of `table`, by vectorizing the formula, as follows:

```
tab(:,4) = 10 + (T(1)-10)*exp(-0.05 * t(1:5/dt:m+1)');
```

The enlarged table should look something like this:

0	25.00	25.00	25.00
5.00	21.61	21.67	21.68
10.00	18.98	19.09	19.10
...			

Note that the numerical solution generated by Equation (10.1) gets more accurate as dt gets smaller. That is because Equation (10.2) (the exact solution) is derived from Equation (10.1) *in the limit* as $dt \to 0$.

Exercise

To complete the analysis of the cooling problem, you may like to superimpose graphs for different values of K and draw them in different colors. It would also be useful to label each graph at some convenient place, e.g., with the label 'K = 0.08'. The command

```
gtext( 'text' )
```

causes an arrow to appear on the graphics screen. You can position the arrow with a mouse or the arrow keys. When you click (or press **Enter**), the specified text appears at that point.

The function gtext takes a string argument. If you want to label with a variable numeric value, use sprintf to convert a number into a string. It works just like fprintf, except that the output goes into a string variable, e.g.,

```
gtext( sprintf('K = %5.3f', K) )
```

- Draw some graphs, on the same set of axes, for different values of K, in different colors, and label them. Fig. 10.1 shows what to aim for.
 Also superimpose the exact solution, given by Equation (10.2).

Plots can also be labeled interactively in plot editing mode or with the Properties Editor.

10.2 FREQUENCIES, BAR CHARTS AND HISTOGRAMS

10.2.1 A random walk

Imagine an ant walking along a straight line, e.g., the x-axis. She starts at $x = 40$. She moves in steps of one unit along the line. Each step is to the left or the right with equal probability. We would like a visual representation of how much time she spends at each position.

Start by running the following script, ant.m:

```
f = zeros(1,100);
x = 40;

for i = 1:1000
  r = rand;
  if r >= 0.5
    x = x + 1;
  else
    x = x - 1;
  end
  if x ~= 0 | x ~= 100
    f(x) = f(x) + 1;
  end
end
```

Now enter the statement bar(f) in the Command Window. You should get a graph which is similar to the one in Fig. 10.2.

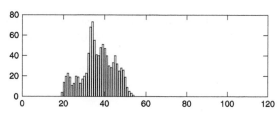

FIGURE 10.2
Random walk frequencies.

Note:

- The function `rand` returns a random number in the range 0–1. If it's greater than 0.5, the ant moves right (x = x + 1), otherwise she moves left (x = x - 1).
- The vector `f` has 100 elements, initially all zero. We define `f(x)` as the number of times the ant lands at position x. Suppose her first step is to the right, so x has the value 41. The statement

 f(x) = f(x) + 1

 then increases the value of `f(41)` to 1, meaning that she has been there once. When she next wanders past this value of x, `f(41)` will be increased to 2, meaning she's been there twice.
 When I ran this script, the final value of `f(41)` was 33—the number of times the ant was there.
- `f(x)` is called a *frequency distribution*, and the graph obtained from `bar(f)` is called a *bar graph*. Each element of `f` is represented by a vertical bar of proportional height. See also `help bar`.
- The script `ant.m` *simulates* the random movement of the ant. If you re-run it, you will get a different bar graph, because `rand` will generate a different sequence of random numbers. Simulation is discussed more fully in Chapter 15.

10.2.2 Histograms

Another helpful way of representing data is with a *histogram*.

As an example, suppose 12 students write a test and obtain the following marks (expressed as percentages), which are assigned to the vector m:

 0 25 29 35 50 55 55 59 72 75 95 100

The statement `hist(m)` will draw a histogram. Try it. The histogram shows the distribution of marks in 10 'bins' (categories) which are equally spaced be-

tween the minimum (0) and maximum (100) marks in the set of data. The number of bins (10 by default) can be specified by a second argument, e.g., hist(m, 25).

To generate the frequencies plotted by hist use the following form (which does not actually draw the histogram):

```
[n x] = hist(m)
```

n is a vector containing the frequencies, viz.,

1	0	2	1	0	4	0	2	0	1	1

For example, there is one mark in the first bin (0–9), none in the second (10–19), two in the third, etc.

The second output vector x contains the midpoints of the bins, such that bar(x, n) plots the histogram. See Help for more details.

Note the subtle difference between a histogram and a bar graph. The values plotted by hist are computed from the distribution of values in a vector, whereas bar generates a bar graph directly from the values themselves.

10.3 SORTING

One of the standard applications of arrays is sorting a list of numbers into, say, ascending order. Although MATLAB has its own sorting function (sort), you may be interested in how sorting algorithms actually work.

10.3.1 Bubble Sort

The basic idea is that the unsorted list is assigned to a vector. The numbers are then ordered by a process which essentially passes through the vector many times, swopping consecutive elements that are in the wrong order, until all the elements are in the right order. Such a process is called a Bubble Sort, because the smaller numbers rise to the top of the list, like bubbles of air in water. (In fact, in the version shown below, the largest number will 'sink' to the bottom of the list after the first pass, which really makes it a 'Lead Ball' sort.) There are many other methods of sorting, such as the Quick Sort, which may be found in most textbooks on computer science. These are generally more efficient than the Bubble Sort, but its advantage is that it is by far the easiest method to program. A structure plan for the Bubble Sort is as follows:

1. Input the list X
2. Set N to the length of X

Table 10.1 Memory during a Bubble Sort.

	1st pass	2nd pass	3rd pass	4th pass
X_1:	27/13	13/9	9/5	5/3
X_2:	13/27/9	9/13/5	5/9/3	3/5
X_3:	9/27/5	5/13/3	3/9	9
X_4:	5/27/3	3/13	13	13
X_5:	3/27	27	27	27
	4 tests	3 tests	2 tests	1 test

3. Repeat $N - 1$ times with counter K:
 Repeat $N - K$ times with counter J:
 If $X_j > X_{j+1}$ then
 Swop the contents of X_j and X_{j+1}
4. Stop since the list X is now sorted.

As an example, consider a list of five numbers: 27, 13, 9, 5 and 3. They are initially input into the vector X. Part of MATLAB's memory for this problem is sketched in Table 10.1. Each column shows the list during each *pass*. A stroke in a row indicates a change in that variable during the pass as the script works down the list. The number of tests ($X_j > X_{j+1}$?) made on each pass is also shown in the table. Work through the table by hand with the structure plan until you understand how the algorithm works.

Sorting algorithms are compared by calculating the number of tests (comparisons) they carry out, since this takes up most of the execution time during the sort. On the Kth pass of the Bubble Sort there are exactly $N - K$ tests, so the total number of tests is

$$1 + 2 + 3 + \cdots + (N - 1) = N(N - 1)/2$$

(approximately $N^2/2$ for large N). For a list of five numbers there are therefore 10 tests, but for 10 numbers there are 45 tests. The computer time needed goes up as the square of the length of the list.

The function M-file `bubble.m` below departs slightly from the structure plan above, which will make $N - 1$ passes, *even if the list is sorted before the last pass*. Since most real lists are partially sorted, it makes sense to check after each pass if any swops were made. If none were, the list must be sorted, so unnecessary (and therefore time-wasting) tests can be eliminated. In the function, the variable `sorted` is used to detect when the list is sorted, and the outer loop is coded instead as a non-deterministic `while` loop. Here it is:

```
function y = bubble( x )
n = length(x);
sorted = 0;                 % flag to detect when sorted
```

```
k = 0;                      % count the passes

while ~sorted
  sorted = 1;               % they could be sorted
  k = k + 1;                % another pass
  for j = 1:n-k             % fewer tests on each pass
    if x(j) > x(j+1)        % are they in order?
      temp = x(j);          % no ...
      x(j) = x(j+1);
      x(j+1) = temp;
      sorted = 0;           % a swop was made
    end
  end
end;

y = x;
```

You can test it on the command line to sort say 20 random numbers as follows:

```
r = rand(1,20);
r = bubble( r );
```

Note how `bubble` changes its input vector.

On my PC `bubble` takes 1.81 seconds to sort 200 random numbers, but 7.31 seconds to sort 400 numbers. This is consistent with the theoretical result obtained above.

10.3.2 MATLAB's `sort`

The built-in MATLAB function `sort` returns two output arguments: the sorted list (in ascending order) and a vector containing the *indexes* used in the sort, i.e., the positions of the sorted numbers in the original list. If the random numbers

```
r =   0.4175    0.6868    0.5890    0.9304    0.8462
```

are sorted with the command

```
[y,i] = sort(r)
```

the output variables are

```
y =   0.4175    0.5890    0.6868    0.8462    0.9304
i =   1      3      2      5      4
```

For example, the index of the second largest number (0.5890) is 3, which is its subscript in the original unsorted list `r`.

As a matter of fact, the built-in functions `max` and `min` also return second output variables giving indexes.

MATLAB's `sort` is *very* fast. My PC takes 2.36 seconds to sort a list of one million random numbers! This is because (a) a Quick Sort is used, and (b) the script has been compiled as a built-in function; this makes for faster code.

10.4 STRUCTURES

Up to now we have seen arrays with only one type of element—all numeric, or all character. A MATLAB *structure* allows you to put different kinds of data in its various *fields*. For example, we can create a structure called `student` with one field for a student's name,

```
student.name = 'Thandi Mangwane';
```

a second field for her student ID number,

```
student.id = 'MNGTHA003';
```

and a third field for all her marks to date,

```
student.marks = [36 49 74];
```

To see the whole structure, enter its name:

```
student

student =
      name: 'Thandi Mangwane'
        id: 'MNGTHA003'
     marks: [36 49 74]
```

Here's how to access her second mark:

```
student.marks(2)
ans =
    49
```

Note the use of the dot to separate the structure name from its fields when creating it and when accessing its fields.

To add further elements to the structure, use subscripts after the structure name:

```
student(2).name = 'Charles Wilson'
student(2).id = 'WLSCHA007'
student(2).marks = [49 98]
```

(the original student, Thandi Mangwane, is now accessed as `student(1)`). Note that field sizes do not have to conform across elements of a structure array: `student(1).marks` has three elements, while `student(2).marks` has only two elements.

The `student` structure now has size 1-by-2: it has two elements, each element being a student with three fields. Once a structure has more than one element MATLAB does not display the contents of each field when you type the structure name at the command line. Instead, it gives the following summary:

```
student

student =
1x2 struct array with fields:
    name
    id
    marks
```

You can also use `fieldnames(student)` to get this information.

A structure array can be preallocated with the `struct` function. See Help.

A structure field can contain any kind of data, even another structure—why ever not? Thus we can create a structure `course` of courses taken by students, where one field is the name of the course, and another field is a `student` structure with information on all the students taking that particular course:

```
course.name = 'MTH101';
course.class = student;
course

course =
    name: 'MTH101'
   class: [1x2 struct]
```

We can set up a second element of `course` for a different class of students:

```
course(2).name = 'PHY102';
course(2).class = ...
```

To see all the courses:

```
course(1:2).name
ans =
MTH101
ans =
PHY102
```

To see all the students in a particular course:

```
course(1).class(1:2).name
ans =
Thandi Mangwane
ans =
Charles Wilson
```

There is a curious function called deal which 'deals inputs to outputs'. You can use it to generate 'comma-separated variable lists' from structure fields:

```
[name1, name2] = deal(course(1).class(1:2).name);
```

(but you don't actually need the commas here . . .).

You can use the rmfield function to remove fields from a structure.

10.5 CELL ARRAYS

A *cell* is the most general data object in MATLAB. A cell may be thought of as a 'data container', which can contain any type of data: numeric arrays, strings, structures, or cells. An array of cells (and they almost always occur in arrays) is called a *cell array*. While you might think a cell sounds the same as a structure, a cell is more general, and there are also notational differences (which are confusing!).

10.5.1 Assigning data to cell arrays

There are a number of ways of assigning data to cell arrays.

- Cell indexing:

```
c(1,1) = {rand(3)};
c(1,2) = {char('Bongani', 'Thandeka')};
c(2,1) = {13};
c(2,2) = {student};

c =
    [3x3 double]    [2x8 char  ]
    [        13]    [1x2 struct]
```

(assuming that the structure `student` created above still exists). Here the round brackets on the left-hand side of the assignments refer in the normal way to the elements of the cell array. What is different are the curly braces on the right. Curly braces indicate the *contents* of a cell; on the right-hand side of an assignment they are technically cell array *constructors* (remember that each element of this array is a cell).

So read the first statement as

'Construct a cell containing `rand(3)` and assign the cell to element `1,1` of the cell array `c`.'

- Content indexing:

```
c{1,1} = rand(3);
c{1,2} = char('Bongani', 'Thandeka');
c{2,1} = 13;
c{2,2} = student;
```

Here the curly braces on the left indicate the *contents* of the cell element at that particular location. So read the first statement as

'The contents of the cell at location `1,1` becomes `rand(3)`.'

- You can use the curly braces to construct an entire cell array in one statement:

```
b = {[1:5], rand(2); student, char('Jason', 'Amy')}
b =

    [1x5 double]    [2x2 double]
    [1x2 struct]    [2x5 char  ]
```

A cell may contain another cell array; nested curly braces may be used to create nested cell arrays.

- The `cell` function enables you to preallocate empty cell arrays, e.g.

```
a = cell(3,2)      % empty 3-by-2 cell array
a =

    []      []
    []      []
    []      []
```

You can then use assignment statements to fill the cells, e.g.

```
a(2,2) = {magic(3)}
```

Note: if you already have a numeric array with a certain name, don't try to create a cell array of the same name by assignment without first clearing the numeric array. If you don't clear the numeric array, MATLAB will generate an error (it will think you are trying to mix cell and numeric syntaxes).

10.5.2 Accessing data in cell arrays

You can access cell contents using content indexing (curly braces):

```
r = c{1,1}
r =
    0.4447    0.9218    0.4057
    0.6154    0.7382    0.9355
    0.7919    0.1763    0.9169
```

To access a subset of a cell's contents you can concatenate curly braces and round brackets if necessary:

```
rnum = c{1,1}(2,3)
rnum =
    0.9355
```

Here the curly braces (content indexing) indicate the contents of the cell array element c(1,1), which is a 3-by-3 numeric matrix. The subscripts (2,3) then indicate the appropriate element within the matrix.

Curly braces may be concatenated to access nested cell arrays.

10.5.3 Using cell arrays

Cell arrays come into their own when you need to access (different types of) data as 'comma-separated variable lists', as the next example demonstrates.

The functions varargin and varargout, which allow a function to have any number of input or output arguments, are none other than cell arrays. The function testvar has a variable number of input arguments, which are doubled into the variable number of output arguments (assuming that the number of output arguments does not exceed the number of input arguments):

```
function [varargout] = testvar(varargin)

for i = 1:length(varargin)
    x(i) = varargin{i};     % unpack the input args
end

for i = 1:nargout           % how many output arguments?
    varargout{i} = 2*x(i);  % pack up the output args
end
```

Command line:

```
[a b c] = testvar(1, 2, 3, 4)
a =
      2
b =
      4
c =
      6
```

When a function is called with the input argument `varargin` MATLAB automatically packs the list of corresponding input arguments into a cell array. You then simply unpack the cell array inside the function using the curly brace content indexing. You similarly pack up the output arguments into the cell array `varargout`.

Note that, should your function have some compulsory input and output arguments, `varargin` and `varargout` must appear at the end of their respective argument lists.

MATLAB has a discussion on when to use cell arrays in the section **Organizing Data in Cell Arrays**, to be found in **MATLAB Help: Programming and Data Types: Structures and Cell Arrays**.

10.5.4 Displaying and visualizing cell arrays

The function `celldisp` recursively displays the contents of a cell array.

The function `cellplot` draws a visualization of a cell array. Fig. 10.3 depicts the contents of the cell array `c`, created above. Non-empty array elements are shaded.

10.6 CLASSES AND OBJECTS

> You'd have to be living face down in a moon crater not to have heard about object-oriented programming. *Tom Swan.*
>
> Object-oriented programming is an exceptionally bad idea which could only have originated in California. *Edsger Dijkstra.*

MATLAB, along with most other modern programming languages, espouses the cause of object-oriented programming, and has all the usual paraphernalia associated with this paradigm: classes, objects, encapsulation, inheritance, operator overloading, etc., etc.

Object-oriented programming is a subject which requires an entire book to do justice to it. If you want to learn about the concepts, there are many excellent books on the subject. If you want to see how MATLAB implements object-oriented programming, consult the online Help: **MATLAB Help: Programming and Data Types: MATLAB Classes and Objects**.

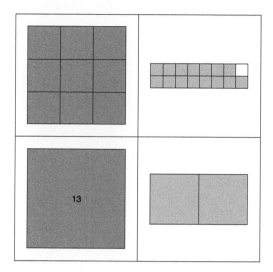

FIGURE 10.3
Result of `cellplot(c)`.

SUMMARY

- A MATLAB structure allows you to store different types of data in its various fields.
- Arrays of structures may be created.
- A cell is the most general data object in MATLAB, and can store any type of data.
- An array of cells is called a cell array—each cell in the array can store different types of data, including other cell arrays.
- A cell array is constructed with curly braces {}.
- The contents of cell in a cell array is accessed with content indexing (curly braces).
- Cell elements in a cell array are accessed in the usual way with round brackets (cell indexing).
- Arguments of functions using a variable number of arguments are packed into cell arrays.
- The function `cellplot` gives a visualization of a cell array.
- MATLAB implements object-oriented programming.

Errors and Pitfalls

THE OBJECTIVE OF THIS CHAPTER IS TO ENABLE YOU TO:

- Begin to recognize and avoid different sorts of errors and pitfalls.

CONTENTS

Syntax errors ... 257
*Incompatible vector
sizes*.................... 257
Name hiding 258
Logic errors 258
Rounding error . 259
Summary 260
Chapter
exercises......... 260

Even experienced programmers seldom get programs to run correctly the first time. In computer jargon, an error in a program is called a *bug*. The story is that a hapless moth short-circuited two thermionic valves in one of the earliest computers. This primeval (charcoaled) 'bug' took days to find. The process of detecting and correcting such errors is therefore called *debugging*. There are a number of different types of errors and pitfalls, some of which are peculiar to MATLAB, and some of which may occur when programming in any language. These are discussed briefly in this chapter.

11.1 SYNTAX ERRORS

Syntax errors are typing errors in MATLAB statements (e.g., `plog` instead of `plot`). They are the most frequent type of error, and are *fatal*: MATLAB stops execution and displays an error message. As MATLAB evolves from one version to the next, error messages improve. Try the following examples to examine the latest error messages:

```
2*(1+3

disp(['the answer is ' num2str(2)]
```

There are many possible syntax errors—you will probably have discovered a few yourself. With experience you will become more adept at spotting your mistakes.

The function `lasterr` returns the last error message generated.

11.1.1 Incompatible vector sizes

Consider the following statements:

Essential MATLAB for Engineers and Scientists. https://doi.org/10.1016/B978-0-08-102997-8.00017-8

```
x = 0:pi/20:3*pi;
y = sin(x);
x = 0:pi/40:3*pi;
plot(x,y)
```

You'll get the error message

```
Error using ==> plot
Vectors must be the same lengths.
```

because you forgot to recalculate y after reducing the x increments. whos reveals the problem:

```
x          1x121       ...
y          1x61        ...
```

11.1.2 Name hiding

Recall that a workspace variable 'hides' a script or function of the same name. The only way to access such a script or function is to clear the offending variable from the workspace.

Furthermore, a MATLAB function hides a script of the same name, e.g., create a script called why.m that displays some junk message, and then type why at the command line.

If you are worried that a variable or script which you are thinking of creating, say blob, may be a MATLAB function, try help blob first.

11.2 LOGIC ERRORS

These are errors in the actual algorithm you are using to solve a problem, and are the most difficult to find; the program runs, but gives the wrong answers! It's even worse if you don't realize the answers are wrong. The following tips might help you to check out the logic.

- Try to run the program for some special cases where you know the answers.
- If you don't know any exact answers, try to use your insight into the problem to check whether the answers seem to be of the right order of magnitude.
- Try working through the program by hand (or use MATLAB's excellent interactive debugging facilities—see Chapter 10) to see if you can spot where things start going wrong.

11.3 ROUNDING ERROR

At times, as we have seen, a program will give numerical answers which we know are wrong. This can also be due to *rounding error*, which results from the finite precision available on the computer, i.e., eight bytes per variable, instead of an infinite number.

Run the following program:

```
x = 0.1;
while x ~= 0.2
   x = x + 0.001;
   fprintf( '%g  %g\n', x, x - 0.2 )
end
```

You will find that you need to crash the program to stop it, i.e., with **Ctrl-break** on a PC. The variable x never has the value 0.2 *exactly*, because of rounding error. In fact, x misses 0.2 by about 8.3×10^{-17}, as can be seen from displaying the value of x - 0.2. It would be better to replace the while clause with

```
while x <= 0.2
```

or, even better, with

```
while abs(x - 0.2) > 1e-6
```

In general, it is always better to test for 'equality' of two non-integer expressions as follows:

```
if abs((a-b)/a) < 1e-6 disp( 'a practically equals b' ), end
```

or

```
if abs((a-b)/b) < 1e-6 ...
```

Note that this equality test is based on the *relative* difference between a and b, rather than on the *absolute* difference.

Rounding error may sometimes be reduced by a mathematical re-arrangement of a formula. Recall yet again the common or garden quadratic equation

$$ax^2 + bx + c = 0,$$

with solutions

$$x_1 = (-b + \sqrt{b^2 - 4ac})/(2a),$$
$$x_2 = (-b - \sqrt{b^2 - 4ac})/(2a).$$

Taking $a = 1$, $b = -10^7$ and $c = 0.001$ gives $x_1 = 10^7$ and $x_2 = 0$. The second root is expressed as the difference between two nearly equal numbers, and considerable significance is lost. However, as you no doubt remember, the product of the two roots is given by c/a. The second root can therefore be expressed as $(c/a)/x_1$. Using this form gives $x_2 = 10^{-10}$, which is more accurate.

SUMMARY

- Syntax errors are mistakes in the construction of MATLAB statements.
- Logical errors are errors in the algorithm used to solve a problem.
- Rounding error occurs because a computer can store numbers only to a finite accuracy.

CHAPTER EXERCISES

11.1 The Newton quotient

$$\frac{f(x+h) - f(x)}{h}$$

may be used to estimate the first derivative $f'(x)$ of a function $f(x)$, if h is 'small'. Write a program to compute the Newton quotient for the function

$$f(x) = x^2$$

at the point $x = 2$ (the exact answer is 4) for values of h starting at 1, and decreasing by a factor of 10 each time (use a `for` loop). The effect of rounding error becomes apparent when h gets 'too small', i.e., less than about 10^{-12}.

11.2 The solution of the set of simultaneous equations

$$ax + by = c$$
$$dx + ey = f$$

(Exercise 3.6) is given by

$$x = (ce - bf)/(ae - bd),$$
$$y = (af - cd)/(ae - bd).$$

If $(ae - bd)$ is small, rounding error may cause quite large inaccuracies in the solution. Consider the system

$$0.2038x + 0.1218y = 0.2014,$$
$$0.4071x + 0.2436y = 0.4038.$$

Show that with four-figure floating point arithmetic the solution obtained is $x = -1$, $y = 3$. This level of accuracy may be simulated in the solution of Exercise 3.6 with some statements like

```
ae = floor( a * e * 1e4 ) / 1e4!
```

and appropriate changes in the coding. The exact solution, obtained without rounding, is $x = -2$, $y = 5$. If the coefficients in the equations are themselves subject to experimental error, the 'solution' of this system using limited accuracy is totally meaningless.

PART

Applications

Part 2 is on applications. Since this is an introductory course, the applications are not extensive. They are illustrative. You ought to recognize that the kinds of problems you actually can solve with MATLAB are much more challenging than the examples provided. However, many of the examples can be a starting point for the development of solutions to more challenging problems. The goal of this part of the text is to begin to scratch the surface of the true power of MATLAB.

Dynamical Systems

THE OBJECTIVE OF THIS CHAPTER IS TO DISCUSS THE IMPORTANCE OF LEARNING TO USE TOOLS LIKE MATLAB.

MATLAB and companion toolboxes provide engineers, scientists, mathematicians, and students of these fields with an environment for technical computing applications. It is much more than a programming language like C, C++ or Fortran. Technical computing includes mathematical computation, analysis, visualization, and algorithm development. The MathWork website describes "The Power of Technical Computing with MATLAB" as follows:

- Whatever the objective of your work—an algorithm, an analysis, a graph, a report, or a software simulation—technical computing with MATLAB lets you work more effectively and efficiently. The flexible MATLAB environment lets you perform advanced analyses, visualize data, and develop algorithms by applying the wealth of functionality available. With its more than 1000 mathematical, statistical, scientific and engineering functions, MATLAB gives you immediate access to high-performance numerical computing. This functionality is extended with interactive graphical capabilities for creating plots, images, surfaces, and volumetric representations.

- The advanced toolbox algorithms enhance MATLAB's functionality in domains such as signal and image processing, data analysis and statistics, mathematical modeling, and control design. Toolboxes are collections of algorithms, written by experts in their fields, that provide application-specific numerical, analysis, and graphical capabilities. By relying on the work of these experts, you can compare and apply a number of techniques without writing code. As with MATLAB algorithms, you can customize and optimize toolbox functions for your project requirements.

CONTENTS

Cantilever beam 267

Electric current 268

Free fall 271

Projectile with friction 280

Summary 284

Exercises 284

In this chapter we are going to examine the application of MATLAB capabilities with four relatively simple problems in engineering science. The problems are the deflection of a cantilever beam subject to a uniform load, a single-loop closed electrical circuit, the free fall problem, and an extension of the projectile problem discussed in Chapter 3. The first problem is on a structural element you investigate in a first course in engineering mechanics. The struc-

Essential MATLAB for Engineers and Scientists. https://doi.org/10.1016/B978-0-08-102997-8.00019-1

tural element is the cantilever beam, which is one of the primary elements of engineered structures, e.g., buildings and bridges. We examine the deflection of this beam with a constant cross section when subject to a uniform distribution of load, e.g., its own weight.

The second problem is an examination of the equation that describes the "flow" of electrical current in a simple closed-loop electrical circuit. You examine this type of problem in a first course in electrical science.

The third problem is the free fall of an object in a gravitational field with constant acceleration, g. This is one of the first problems you examine in a first course in physics. We examine the effect of friction on the free-fall problem and, hence, learn that with friction (i.e., air resistance) the object can reach a terminal velocity.

The fourth problem is an extension of the projectile that takes into account air resistance (you learn about air resistance in your first course in fluid mechanics). By an examination of this problem, you will learn why golfers do not hit the ball from the tee at 45 degrees from the horizontal (which is the optimum angle for the furthest distance of travel of a projectile launched in frictionless air).

Before we begin, you must keep in mind that this is *not* a book on engineering or science; it is an introduction to the essential elements of MATLAB, a powerful technical analysis tool. In this and subsequent chapters you will scratch the surface of its power by solving some relatively simple problems that you have encountered or will encounter early in your science or engineering education. Keep in mind that MATLAB is a tool that can be used for a number of very productive purposes. One purpose is to learn how to write programs to investigate topics you confront in the classroom. You can extend your knowledge of MATLAB by self-learning to apply the utilities and any of the toolboxes that are available to you to solve technical problems. This tool requires a commitment from you to continue to discover and to use the power of the tool as part of your continuing education. Developing high-quality computer programs within the MATLAB environment in order to solve technical problems requires that you use the utilities available. If you need to write your own code you should, if at all possible, include the utilization of generic solvers of particular types of equations that are available in MATLAB. Of course, to do this you need to learn more advanced mathematics, science and engineering. You will certainly have the opportunity to do this after freshman year.

As a student majoring in science and engineering you will take at least five semesters of mathematics. The fifth course is usually titled Advanced Calculus, Advanced Engineering Mathematics, Applied Mathematics, or Methods of Mathematical Physics. Topics covered in this course are typically ordinary differential equations, systems of differential equations (including nonlinear systems), linear algebra (including matrix and vector algebra), vector calculus,

Fourier analysis, Laplace transforms, partial differential equations, complex analysis, numerical methods, optimization, probability and statistics. By clicking the question mark (?) near the center of the upper-most tool bar in the MATLAB desk top, as you already know, opens the help browser. Scan the topics and find utilities and examples of approaches to solve problems in all of these subjects and more.

The Control Systems Toolbox is an example of one of the numerous engineering and scientific toolboxes that have been developed to advance the art of doing science and engineering. SIMULINK and Symbolics are also very important toolboxes. The latter two are included with your "student version" of MATLAB. One of the big advantages of SIMULINK, in addition to being a graphical algorithm-development and programming environment, is that it is useful when using the computer to measure, record and access data from laboratory experiments or data from industrial-process monitoring systems. Symbolics is very useful because it allows you to do symbolic mathematics, e.g., integrating functions like the ones you integrate in your study of the calculus. All of the mathematical subjects investigated by you, as a science or engineering major, are applied in your science and engineering courses. They are the same techniques that you will apply as a professional (directly or indirectly depending on the software tools available within the organization that you are employed). Thus, you can use your own copy of MATLAB to help you understand the mathematical and software tools that you are expected to apply in the classroom as well as on the job.

12.1 CANTILEVER BEAM

In this section we want to examine the problem of the cantilever beam. The beam and its deflection under load is illustrated in Fig. 12.1 below, which is generated by the script *m-file* created to solve the problem posed next. Many structural mechanics formulas are available in *Formulas for Stress and Strain*, Fifth Edition by Raymond J. Roark and Warren C. Young, published by McGraw-Hill Book Company (1982). For a uniformly loaded span of a cantilever beam attached to a wall at $x = 0$ with the free end at $x = L$, the formula for the vertical displacement from $y = 0$ under the loaded condition with y the coordinate in the direction opposite that of the load can be written as follows:

$$Y = \frac{y\,24EI}{wL^4} = -\left(X^4 - 4X^3 + 6X^2\right),$$

where $X = x/L$, E is a material property known as the modulus of elasticity, I is a geometric property of the cross-section of the beam known as the moment of inertia, L is the length of the beam extending from the wall from which it is mounted, and w is the load per unit width of the beam (this is a two-dimensional analysis). The formula was put into dimensionless form to

answer the following question: What is the shape of the deflection curve when the beam is in its loaded condition and how does it compare with its unloaded perfectly horizontal orientation? The answer will be provided graphically. This is easily solved by using the following script:

```
%
% The deflection of a cantilever beam under a uniform load
% Script by D.T.V. ............. September 2006, May 2016
% Revised 2018
% Step 1: Select the distribution of X's from 0 to 1
% where the deflections to be plotted are to be determined.
%
X = 0:.01:1;
%
% Step 2: Compute the deflections Y at eaxh X. Note that YE is
% the unloaded position of all points on the beam.
%
Y = - ( X.^4 - 4 * X.^3 + 6 * X.^2 );
YE = 0;
%
% Step 3: Plot the results to illustrate the shape of the
% deflected beam.
%
plot([0 1],[0 0],'--',X,Y,'LineWidth',2)
axis([0,1.5,-4, 1]),title('Deflection of a cantilever beam')
xlabel('X'),ylabel('Y')
legend('Unloaded cantilever beam','Uniformly loaded beam')
%
% Step 4: Stop
```

Again, the results are plotted in Fig. 12.1. It looks like the beam is deflected tremendously. The actual deflection is not so dramatic once the actual material properties are substituted to determine y and x in, say, meters. The scaling, i.e., re-writing the equation in dimensionless form in terms of the unitless quantities Y and X, provides insight into the shape of the curve. In addition, the shape is independent of the material as long as it is a material with uniform properties and the geometry has a uniform cross sectional area, e.g., rectangular with height h and width b that are independent of the distance along the span (or length) L.

12.2 ELECTRIC CURRENT

When you investigate electrical science you investigate circuits with a variety of components. In this section we will solve the governing equation to examine

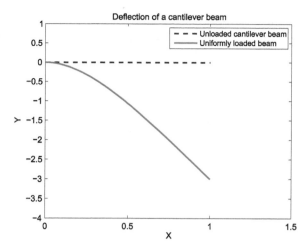

FIGURE 12.1
The vertical deflection of a uniformly loaded, cantilever beam.

the dynamics of a single, closed loop electrical circuit. The loop contains a voltage source, V (e.g., a battery), a resistor, R (i.e., an energy dissipation device), an inductor, L (i.e., an energy storage device), and a switch which is instantaneously closed at time $t = 0$. From Kirchoff's law, as described in the book by Ralph J. Smith entitled *Circuits, Devices, and Systems*, published by John Wiley & Sons, Inc. (1967), the equation describing the response of this system from an initial state of zero current is as follows:

$$L\frac{di}{dt} + Ri = V,$$

where i is the current. At $t = 0$ the switch is engaged to close the circuit and initiate the current. At this instant of time the voltage is applied to the resistor and inductor (which are connected in series) instantaneously. The equation describes the value of i as a function of time after the switch is engaged. Hence, for the present purpose, we want to solve this equation to determine i versus t graphically. Rearranging this equation, we get

$$\frac{di}{dt} + \frac{R}{L}i = \frac{V}{L}.$$

The solution, by inspection (another method that you learn when you take differential equations), is

$$i = \frac{V}{R}\left(1 - e^{-\frac{R}{L}t}\right).$$

This solution is checked with the following script:

```
%
% Script to check the solution to the soverning
% equation for a simple circuit, i.e., to check
% that
%       i = (V/R) * (1 -  exp(-R*t/L))
%
% is the solution to the following ODE
%
%       di/dt + (R/L) * i - V/L = 0
%
% Step 1: We will use the Symbolics tools; hence,
% define the symbols as follows
%
        syms i V R L t
%
% Step 2: Construct the solution for i
%
        i = (V/R) * ( 1 - exp(-R*t/L) );
%
% Step 3: Find the derivative of i
%
        didt = diff(i,t);
%
% Step 4: Sum the terms in ODE
%
        didt + (R/L) * i - V/L;
%
% Step 5: Is the answer ZERO?
%
        simple(ans)
%
% Step 6: What is i at t = 0?
%
subs(i,t,0)
%
% REMARK: Both answers are zero; hence,
%         the solution is correct and the
%         initial condition is correct.
%
% Step 7: To illustrate the behavior of the
%         current, plot i vs. t for V/R = 1
%         and R/L = 1. The curve illustrates
%         the fact that the current approaches
%         i = V/R exponentially.
%
```

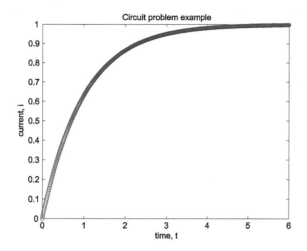

FIGURE 12.2
Exponential approach to steady current condition of a simple RL circuit with an instantaneously applied constant voltage.

```
        V = 1; R = 1; L = 1;
        t = 0 : 0.01 : 6;
        i = (V/R) * ( 1 - exp(-R.*t/L) );
        plot(t,i,'ro'), title(Circuit problem example)
xlabel('time, t'),ylabel('current, i')
%
```

When you run this script you will prove that the solution given above is correct. The figure that illustrates the solution to the problem posed is reproduced in Fig. 12.2. This concludes this exercise related to electrical circuit theory that you will learn more about in your course on electrical science.

12.3 FREE FALL

In this section we are going to use MATLAB to not only investigate the problem of free fall with friction (or air resistance), we are going to use MATLAB to check the theoretical results found in the literature. This example deals with the application of MATLAB for computing and graphically presenting information for analysis of the problem of free fall. This investigation is to examine the effect of air resistance on the distance of free fall in 5 seconds from a location $y = 0$, where the object is initially at rest (y is in the direction of gravity). Hence, we want to determine the distance, $y = L$ that an object falls from a state of rest with and without air resistance. In the book entitled *Introduction to Theoretical Mechanics* by R.A. Becker, that was published by McGraw-Hill Book Company (1954), the equations for free fall are given. The three cases of interest are as follows:

1. Without air resistance:

$$a = \frac{d^2y}{dt^2} = g, \quad v = \frac{dy}{dt} = gt, \quad y = \frac{1}{2}gt^2,$$

where a is the acceleration of the object, v is its speed and y is the distance of its free fall from the start of motion at $t = 0$.

2. With resistance proportional to the linear power of velocity:

$$a = \frac{d^2y}{dt^2} = g - k\frac{dy}{dt}, \quad v = \frac{dy}{dt} = \frac{g}{k}\left(1 - e^{-kt}\right), \quad y = \frac{g}{k}t - \frac{g}{k^2}\left(1 - e^{-kt}\right).$$

3. With resistance proportional to the second power of velocity:

$$a = \frac{d^2y}{dt^2} = g - k\left(\frac{dy}{dt}\right)^2, \quad v = \frac{dy}{dt} = \frac{1}{2}\frac{g}{k}\tanh\left(\frac{gk}{2}t\right),$$

$$y = \frac{1}{k}\log_e\left[\cosh\left(gkt/2\right)\right].$$

For all three cases the initial condition is $y = v = 0$ at $t = 0$. (You will learn more about ordinary differential equations in your third semester of mathematics. In addition, in a first course in fluid mechanics you will learn about some of the details about air resistance. In particular, for air resistance associated with "laminar flow" problems, case 2 applies. In many practical situations it is case 3 that is usually the case of interest; this is the case where "turbulent flow" is important.)

Let us consider the following problem: Let us assume the above equations are correct. Note that they are written for a unit of mass. Let us also assume $g = 9.81$ m/s^2 and $k = 0.2$ for the air-drag parameter. Finally, let us answer the following:

(a) What is y in meters for $t = 5$ seconds of free fall for the three cases.
(b) What are the terminal speeds in meters per second for cases 2 and 3 for the specified air-drag parameter?
(c) Is the terminal speed reached at $t = 5$ seconds?

Note:

The terminal speeds are g/k and $(g/k)/2$ for cases 2 and 3, respectively, where the terminal speed is the time independent or steady speed reached after a sufficient distance of free fall. It is the speed at which the gravitational force balances the air resistance force. From Part I, the Essentials, we learned that MATLAB is quite useful in this type of problem because it has the capability of computing the elementary functions in the formulas above.

As part of providing an answer to the questions raised, we want to examine the distance of free fall within the 5 seconds of flight time with the equations for y

given above. Let us examine the influence of air resistance on free fall graphically; thus, we need to plot the distance y versus t from $t = 0$ to $t = 5$ seconds. We shall plot all three curves on one figure for direct comparison. We will show that for a short period of time (significantly less than the 5 seconds) after the onset of motion all three curves are on top of each other. We will also show that at $t = 5$ seconds the distances of free fall are quite different.

The steps in the structure plan and the MATLAB code are as follows:

```
%
%       Free fall analysis (saved as FFall.m):
%         Comparison of exact solutions of free
%         fall with zero, linear and quadratic
%         friction for t = 0 to 5 seconds.
%
% Script by D. T. V. ......... September 2006.
% Revised by D.T.V.   ......... 2008/2016/2018.
%
% Step 1: Specify constants
%
% Friction coefficient provided in the problem statement.
k = 0.2;
% Acceleration of gravity in m/s/s.
g = 9.81;
%
% Step 2: Selection of time steps for computing solutions
%
dt = .01;
%
% Step 3: Set initial condition (the same for all cases)
%
t(1) = 0.; v(1) = 0.; y(1) = 0.;
%
t = 0:dt:5;
%
% Step 4: Compute exact solutions at each time step
%             from t = 0 to 5.
%
%     (a) Without friction:
%
            v = g * t;
            y = g * t.^2 * 0.5;
%
%     (b) Linear friction
%
```

```
            velf = (g/k) * (1. - exp(-k*t));
            yelf = (g/k) * t - (g/(k^2)) * (1.-exp(-k*t));
%
%    (c) Quadratic friction
%
            veqf = sqrt(g/k) * tanh( sqrt(g*k) * t);
            yeqf = (1/k) * log(cosh( sqrt(g*k) * t) );
%
% Step 5: Computation of the terminal speeds
%         (cases with friction)
%
            velfT = g/k;
            veqfT = sqrt(g/k);
%
% Step 6: Graphical comparison
%
 plot(t,y,t,yelf,t,yeqf)
 title('Fig 1. Comparison of results')
 xlabel(' Time, t')
 ylabel(' Distance, y ')
 figure
 plot(t,v,t,velf,t,veqf)
 title('Fig. 2. Comparison of results')
 xlabel(' Time, t')
 ylabel(' Speed, v ')
%
% Step 7: Comparison of distance and speed at t = 5
%
disp(' ');
fprintf(' y(t) = %f, yelf(t) = %f, yeqf(t) = %f at t = %f\n',...
        y(501),yelf(501),yeqf(501),t(501))
disp(' ');
fprintf(' v(t) = %f, velf(t) = %f, veqf(t) = %f at t = %f\n',...
        y(501),yelf(501),yeqf(501),t(501))
%
% Step 8: Comparison of terminal velocity
%
disp(' ');
fprintf('   velfT = %f, veqfT = %f\n',...
   velfT,veqfT)
%
% Step 9: Stop
%
```

FIGURE 12.3
Comparison of free-fall distance: Top curve is for no friction, middle curve is for linear friction and the bottom curve is for quadratic friction.

The command window execution of the file above (named FFALL.m) gives the comparisons in Figs. 12.3 and 12.4 and the printed results as follows:

```
>> FFall

y(t) = 122.625000, yelf(t) = 90.222433, yeqf(t) = 31.552121 at t = 5.000000
v(t) = 49.050000, velf(t) = 31.005513, veqf(t) = 7.003559 at t = 5.000000
velfT = 49.050000, veqfT = 7.003571
```

The figures illustrate, as we may have expected, that for no friction the object falls the furthest distance. The case with quadratic friction reaches terminal velocity well within the five seconds examined. The linear friction case did not reach terminal speed, yet it is moving at a slower velocity as compared with the no friction case. Keep in mind that a unit mass object is falling with a friction coefficient $k = 0.2$. The same k was used for both the second and third cases. Within the first half second from the time of release the three curves are not distinguishable illustrating the fact that it takes a little time before the friction effects are felt. At the time of 5 seconds after release, the speed and the distance fallen are quite different. It is not surprising that quadratic friction slows the object quicker because the air resistance (or friction) is proportional to the speed squared, which is significantly larger than speed to the first power (as it is in the linear-friction case).

The above analysis was based on the exact solutions of the free-fall problem. Let use now use the symbolic tools to check the theoretical results found in the literature. The case that we will examine is Case 2, the linear friction case. The following script was implemented in the Command Window. The responses of MATLAB are also reproduced.

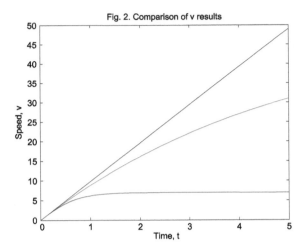

FIGURE 12.4

Comparison of free-fall speed: Top curve is for no friction, middle curve is for linear friction and the bottom curve is for quadratic friction.

```
%
% The formula for the distance of free fall
% of an object from rest with linear friction
% is as follows:
%
% y = (g / k) * t - (g/k^2) * [ 1 - exp^(- k t)].
%
% To check the theory, we want to differentiate
% this twice to determine the formulas for velocity
% v and acceleration a, respectively. The results
% should match the published results.
%
% Step 1: Define the symblic variables
%
syms g k t y
%
% Step 2: Write the formula for y = f(t)
%
y = (g/k) * t - (g/k^2) * ( 1 - exp(-k * t));
%
% Step 3: Determine the velocity
%
v = diff(y,t);
%
% Step 4: Determine the acceleration
%
```

```
a = diff(v,t);
%
% Step 5: Print the v and a formulas and compare with
% the published results
%
v, a
%
% Step 6: Determine a from published formula and v.
%
a2 = g - k * v;
% Step 7: Simplify to somplest form
a2 = simple(a2)
%
% Step 6: Stop. REMARK: Results compare exactly. The
% results printed in the command window after executing
% this script are as follows:
%    v = g/k-g/k*exp(-k*t)
%    a  = g*exp(-k*t)
%    a2 = g/exp(k*t)
% These results verify the conclusion that the published
% formulas are correct.
%
```

Let us next consider an approximate method for solving the linear-friction case. This procedure is something you could have implemented if you didn't have the exact solution. More on numerical methods is provided in the last chapter of this book.

The equations for free-fall acceleration and velocity are differential equations. The following approximate analysis of a differential equation is called finite-differences. Let us consider the problem of free fall with linear friction (i.e., air drag that is linearly proportional to the speed of free fall). The formula (or equation) for this problem and its exact solution are given above. In the following analysis a script to solve this problem by the approximate method described below was written and executed for the same interval of time. This is described next. Then the approximate solution is compared with the exact solution graphically. For a unit mass, the formula that describes the velocity of free fall from rest with air resistance proportional to the linear power of velocity can be written as follows:

$$\frac{dv}{dt} = g - kv,$$

Let us approximate this equation by inverting the fundamental theorem of differential calculus, i.e., let us re-write it in terms of the definition of what a

derivative is before the appropriate limit is taken. What is meant by this is to write this equation for a small interval of time $\Delta t = t(n+1) - t(n)$ as follows:

$$\frac{dv}{dt} \approx \frac{v(n+1) - v(n)}{\Delta t} = g - k\left(\frac{v(n+1) + v(n)}{2}\right),$$

where the value identified by the integer n is the value of v at the beginning of the time interval, i.e., $v(n)$ at $t(n)$. The value identified by $n+1$ is the value of v at the end of the time interval, i.e., $v(n+1)$ at $t(n+1)$. This is an initial-value problem; hence, the value of v is known at $t(n)$. Knowing $v(n)$ at $t(n)$, and by specifying $t(n+1)$, the value $v(n+1)$ can be calculated by solving the finite-difference equation for $v(n+1)$ given above. The solution will depend on the size of Δt. This formula gives us $v(n+1)$. We next need to solve for $y(n+1)$ from the definition of v (viz., an approximate form of dy/dt). The average value of v over the interval of time, Δt, can be written in terms of y as follows:

$$\frac{dy}{dt} \approx \frac{y(n+1) - y(n)}{\Delta t} = \frac{v(n+1) + v(n)}{2}.$$

Rearranging this equation, we get a formula for $y(n+1)$ in terms of $v(n+1)$, $v(n)$, $y(n)$ and Δt. Given the initial condition, i.e., $y(n)$ and $v(n)$ at $t(n)$, we compute $v(n+1)$ with the first difference equation followed by computing $y(n+1)$ with the last equation. Note *that the new values become the initial condition for the next time step.* This procedure is done repeatedly until the overall time duration of interest is reached (in this case this is $t = 5$). The script file to implement this procedure is given next. The $m - file$ is executed and the results plotted on the same graph with the exact solution. As shown, the comparison is remarkable. This is not always the case when approximate methods are used. The comparison is, of course, encouraging because we do not usually have exact results and need to resort to numerical approximations in our work.

The script file applied to examine the difference between the approximate method and the exact result is as follows:

```
%
% Approximate and exact solution comparison of
% free fall with linear friction for t = 0 to 5.
%
% Script by D. T. V. .......... September 2006.
% Revised by D.T.V. .......... 2008/2016/2018.
%
% Step 1: Specified constants
%
k = 0.2;
g = 9.81;
```

```
%
% Step 2: Selection of time step for approximate solution
%               dt = t(n+1) - t(n)
dt = 0.01;
%
% Step 3: Initial condition
%
t(1) = 0.;
v(1) = 0.;
y(1) = 0.;
%
% Step 3: Sequential implementation of the approximate
% solution method by repeating the procedure 500 times
% (using a for loop).
%
for n = 1:1:500
    t(n+1) = t(n) + dt;
    v(n+1) = (v(n) + dt * (g-0.5*k*v(n)) )/(1.+dt*0.5*k);
    y(n+1) = y(n) + dt * 0.5 * (v(n+1) + v(n));
end
%
% Step 4: Exact solution over the same interval of time:
%
    ye = (g/k) * t - (g/(k^2)) * (1.-exp(-k*t));
%
% Step 5: Graphical comparison:
%
plot(t,y,'o',t,ye)
title('Comparison of numerics w/ exact results')
xlabel(' Time, t')
ylabel(' Distance of free fall from rest, y')
%
% Step 6: Comparison of distance at t=5
%
disp(' ');
fprintf(' y(t) = %f, ye(t) = %f at t = %f \n',...
        y(501),ye(501),t(501))
%
% Step 7: End of script by DTV.
%
```

In summary, the examination of the free fall problem illustrated the application and the checking of formulas reported in the literature. In addition, an approximate method was applied to solve the same problem to illustrate

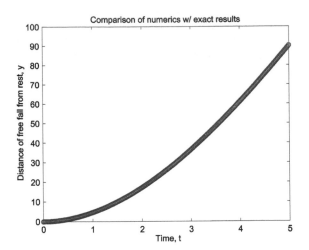

FIGURE 12.5
Comparison of free-fall distance. speed: Exact is the line, circles are for the finite-difference approximate solution.

the application of a method that would be necessary if exact formulas were not found. Finally, the approximate method that was applied can certainly be improved by utilizing the ordinary differential equation solvers available in MATLAB. Examples of this type of procedure are given in the last chapter. Other examples are in the help manuals in MATLAB that are found via the question mark (?) in the toolbar on the desktop window system (as already mentioned). (See Fig. 12.5.)

12.4 PROJECTILE WITH FRICTION

Let us examine the projectile problem again. In this case let us examine the effect of air resistance on the flight of a projectile like a golf ball. For a unit mass, the formulas that describe the trajectory of a projectile in a gravitational field, g, with air resistance in the opposite direction of motion (that is proportional to the speed of the projectile in the direction of motion squared) are as follows:

$u = dx/dt,$

$v = dy/dt,$

$du/dt = -ku * \sqrt{u^2 + v^2},$

$dv/dt = -kv * \sqrt{u^2 + v^2} - g.$

$dv/dt = -kv * \sqrt{u^2 + v^2} - g.$

The location of the projectile (e.g., a golf ball) is initially at $x = 0$, $y = 0$. It is launched at a speed V_s in the direction of angle θ as measured from the hori-

zontal. The coordinate *x* is in the horizontal direction parallel to the ground. The coordinate *y* is in the direction perpendicular to *x* pointing towards the sky. Hence, the gravitational acceleration is in the negative *y* direction. For a given launch speed and direction, we wish to estimate the range (or the distance) the ball travels in the *x* direction when it first hits the ground. To do this we shall approximate the four equations in a similar way that we treated the free falling object problem in the last section.

The solution method is given in detail in the script provided at the end of this paragraph. It was saved with the name golf.m.

```
%
% "The Golf ball problem"
% Numerical computation of the trajectory of a
% projectile launched at an angle theta with
% a specified launch speed. They are:
%  theta = launch angle in degrees.
%  Vs = launch speed.
%
%  Script by D. T. V. ......... September 2006.
%  Revised by D.T.V.  ........ 2008/2016/2018.
%
% Equations of motion:
%  u = dx/dt.  v = dy/dt. g is in the opposite
%  direction of y. x = y = 0 is the location of
%  the tee. k is the coefficient of air drag. It
%  is assumed to be constant. Friction is assumed
%  to be proportional to the speed of the ball
%  squared and it acts in the opposite direction
%  of the direction of motion of the ball. The
%  components of acceleration are thus:
%  du/dt = - [k (u^2 + v^2) * u/sqrt(u^2+v^2)].
%  dv/dt = - [k (u^2 + v^2) * v/sqrt(u^2+v^2)] - g.
%
% INPUT DATA
% Specified constants:
    k = 0.02;
    g = 9.81;
    dt = 0.01;
%
% Input the initial condition:
%
theta = input(' Initial angle of launch: ')
the = theta * pi/180.;
Vs = input('  Initial speed of launch:  ')
```

```
u(1) = Vs * cos(the);
v(1) = Vs * sin(the);
% Launch pad location:
x(1) = 0.;
y(1) = 0.;
%
% Compute approximate solution of the trajectory
% of flight.
% Repeat up to 6000 time, i.e., until ball hits
% the ground.
%
for n=1:1:6000;
    u(n+1) = u(n) ...
      - dt * (k * sqrt(u(n)^2+v(n)^2) * u(n));
    v(n+1) = v(n) ...
      - dt * (k * sqrt(u(n)^2+v(n)^2) * v(n) + g);
    x(n+1) = x(n) + u(n) * dt;
    y(n+1) = y(n) + v(n) * dt;
% Determination of when the object hits ground:
  if y(n+1) < 0
      slope = (y(n+1) - y(n))/(x(n+1) - x(n));
      b = y(n) - slope * x(n);
      xhit = - b/slope;
    plot(x,y)
    fprintf(' The length of the shot =  %5.2f \n', xhit)
  end
% Once object hits terminate the computations with a break:
     if y(n+1) < 0; break; end
end
%
% Graphical presentation of the results:
%
if y(n+1) > 0
    plot(x,y)
end
% ------ End of golf-ball script by DTV
```

What is the optimum launch angle for $k = 0.02$? The answer to this question is described next for the launch speed of 100. Note that the optimum angle is the angle which gives the longest distance of travel (or greatest range). To compute the optimum angle the following script was executed after the lines with the input statements in the golf.m scripted were commented out by typing % at the beginning of the two lines with the input commands. The two lines in question were commented out as follows:

```
%
% In golf.m the following alterations were made
% prior to running this script!!!!!!
%
% % theta = input(' Initial angle of launch: ')
% % Vs = input('  Initial speed of launch:  ')
%
%  This script then finds the optimum angle for k = 0.2
%  to compare with the zero friction case, which we know
%  has the optimum launch angle of 45 degrees.
%
% Script by D. T. V. ......... September 2006.
% Revised by D.T.V.  ........ 2008/2016/2018.
%
% Consider launch angles from 1 to 45 degrees
%
th = 1:1:45;
vs = 100;          % Specified launch speed.
%
% Execute the modified golf.m file 45 times and save
% results for each executoion of golf.m
%
for i=1:45
    theta = th(i)
    golf % Execution of modified golf.m script.
    xh(i) = xhit
    thxh(i) = theta
end
% Find the maximum distance and the correponding index
[xmh,n] = max(xh)
% Determine the angle that the maximum distance occured.
opt_angle = thxh(n)
% Display the results
disp(' optimum angle ')
disp( opt_angle )
% REMARK: For this case the result is 30 degrees.
% End of script
```

The optimum angle of launch for the case of nonlinear friction was computed and found to be equal to 30 degrees. Without friction it is 45 degrees. Hence, it is not surprising that golfers launch their best drives at angles significantly less than 45 degrees.

SUMMARY

In this chapter we examined three problems using some of the utilities in MATLAB. We did the following:

- We determined the shape of the deflection of a cantilever beam based on a published formula in the engineering literature. Thus, we could use the capability of MATLAB to do arithmetic with polynomials.
- We examined the effect of friction on the free fall problem. In addition to computing the distance and speed of free fall based on published formulas, we checked the formulas with the symbolics tools and we also solved the problem by an approximate method to illustrate the range of possibilities at your disposal to solve technical problems.
- We examined the projectile problem subjected to air resistance. We did this by applying the same type of approximate method as we applied in the free fall problem. We found that the optimum angle of launch with friction taken into account is less than 45 degrees (the frictionless value).

EXERCISES

12.1 Reproduce the script for the exact solutions for the free fall problem (called FFall.m) and execute it for a range of friction coefficients, e.g., $k = 0.1$ and 0.3.

12.2 Use the symbolics tools in the same way they were used to check the formulas for the linear-friction case of free fall, to check the other two cases.

12.3 Reproduce the golf.m (projectile script) and look at the effect of varying the friction coefficient k. Look at $k = 0.01$ and 0.03. What influence does friction have on the optimum angle for a launch speed of 100?

Simulation

THE OBJECTIVE OF THIS CHAPTER IS TO INTRODUCE YOU TO:

- Simulation of "real-life" events.

CONTENTS

Random number
generation 285
Seeding rand 286

Spinning coins .. 286

Rolling dice 287

Bacteria division 288

A random walk . 288

Traffic flow 290

Normal
(Gaussian)
random
numbers 293

Summary 293

Exercises 294

Simulation is an area of application where computers have come into their own. A simulation is a *computer experiment* which mirrors some aspect of the real world that appears to be based on random processes, or is too complicated to understand properly. (Whether events can be really random is actually a philosophical or theological question.) Some examples are: radio-active decay, rolling dice, bacteria division and traffic flow. The essence of a simulation program is that the programmer is unable to predict beforehand exactly what the outcome of the program will be, which is true to the event being simulated. For example, when you spin a coin, you do not know for sure what the result will be.

13.1 RANDOM NUMBER GENERATION

Random events are easily simulated in MATLAB with the function `rand`, which we have briefly encountered already. By default, `rand` returns a *uniformly distributed pseudo-random* number in the range $0 \leq$ rand < 1. (A computer cannot generate truly random numbers, but they can be practically unpredictable.) `rand` can also generate row or column vectors, e.g., `rand(1,5)` returns a row vector of five random numbers (1 row, 5 column) as shown:

```
0.9501    0.2311    0.6068    0.4860    0.8913
```

If you generate more random numbers during the same MATLAB session, you will get a different sequence each time, as you would expect. However, each time you start a MATLAB session, the random number sequence begins at the same place (0.9501), and continues in the same way. This is not true to

285

life, as every gambler knows. To produce a different sequence each time you start a session, rand can be initially *seeded* in a different way each time you start.

13.1.1 Seeding rand

The random number generator rand can be seeded with the statement

```
rand('state', n)
```

where n is any integer. (By default, n is set to 0 when a MATLAB session starts.) This is useful if you want to generate the *same* random sequence every time a script runs, e.g., in order to debug it properly. Note that this statement does not generate any random numbers; it only initializes the generator.

However, you can also arrange for n to be different each time you start MATLAB by using the system time. The function clock returns the date and time in a six-element vector with seconds to two decimal places, so the expression sum(100*clock) never has the same value (well, hardly ever). You can use it to seed rand as follows:

```
≫rand('state', sum(100*clock))
≫rand(1,7)
ans =
   0.3637  0.2736  0.9910  0.3550  0.8501  0.0911  0.4493
≫rand('state', sum(100*clock))
≫rand(1,7)
ans =
   0.9309  0.2064  0.7707  0.7644  0.2286  0.7722  0.5315
```

Theoretically rand can generate over 2^{1492} numbers before repeating itself.

13.2 SPINNING COINS

When a fair (unbiased) coin is spun, the probability of getting heads or tails is 0.5 (50%). Since a value returned by rand is equally likely to anywhere in the interval (0, 1) we can represent heads, say, with a value less than 0.5, and tails otherwise.

Suppose an experiment calls for a coin to be spun 50 times, and the results recorded. In real life you may need to repeat such an experiment a number of

times; this is where computer simulation is handy. The following script simulates spinning a coin 50 times:

```
for i = 1:50
  r = rand;
  if r < 0.5
    fprintf( 'H' )
  else
    fprintf( 'T' )
  end
end
fprintf( '\n' )        % newline
```

Here is the output from two sample runs:

```
THHTTHHHHTTTTTHTHTTTHHTHTTTHHTTTTHTTHHTHHHHHHTTHTT
THTHHHTHTHHTTHTHTTTHHTTTTTTTHHHTTTHTHHTHHHHTTHTHTT
```

Note that it should be impossible in principle to tell from the output alone whether the experiment was simulated or real (if the random number generator is sufficiently random).

Can you see why it would be wrong to code the `if` part of the coin simulation like this:

```
if rand < 0.5 fprintf( 'H' ), end
if rand >= 0.5 fprintf( 'T' ), end
```

The basic principle is that `rand` should be called only *once* for each 'event' being simulated. Here the single event is spinning a coin, but `rand` is called twice. Also, since two different random numbers are generated, it is quite possible that *both* logical expressions will be true, in which case 'H' and 'T' will both be displayed for the same coin!

13.3 ROLLING DICE

When a fair dice is rolled, the number uppermost is equally likely to be any integer from 1 to 6. We saw in **Counting random numbers** (Chapter 5) how to use `rand` to simulate this. The following statement generates a vector with 10 random integers in the range 1–6:

```
d = floor( 6 * rand(1,10) + 1 )
```

Here are the results of two such simulations:

| 2 | 1 | 5 | 5 | 6 | 3 | 4 | 5 | 1 | 1 |
| 4 | 5 | 1 | 3 | 1 | 3 | 5 | 4 | 6 | 6 |

We can do statistics on our simulated experiment, just as if it were a real one. For example, we could estimate the mean of the number obtained when the dice is rolled 100 times, and the probability of getting a six, say.

13.4 BACTERIA DIVISION

If a fair coin is spun, or a fair dice is rolled, the different events (e.g., getting 'heads', or a 6) happen with equal likelihood. Suppose, however, that a certain type of bacteria divides (into two) in a given time interval with a probability of 0.75 (75%), and that if it does not divide, it dies. Since a value generated by rand is equally likely to be anywhere between 0 and 1, the chances of it being *less than* 0.75 are precisely 75%. We can therefore simulate this situation as follows:

```
r = rand;
if r < 0.75
  disp( 'I am now we' )
else
  disp( 'I am no more' )
end
```

Again, the basic principle is that one random number should be generated for each event being simulated. The single event here is the bacterium's life history over the time interval.

13.5 A RANDOM WALK

A seriously short-sighted sailor has lost his contact lenses returning from a symphony concert, and has to negotiate a jetty to get to his ship. The jetty is 50 paces long and 20 wide. He is in the middle of the jetty at the quay-end, point-ing toward the ship. Suppose at every step he has a 60% chance of stumbling blindly toward the ship, but a 20% chance of lurching to the left or right (he manages to be always facing the ship). If he reaches the ship-end of the jetty, he is hauled aboard by waiting mates.

The problem is to simulate his progress along the jetty, and to estimate his chances of getting to the ship without falling into the sea. To do this correctly, we must simulate one *random walk* along the jetty, find out whether or not he reaches the ship, and then repeat this simulation 1000 times, say (if we have a fast enough computer!). The proportion of simulations that end with the sailor safely in the ship will be an estimate of his chances of making it to the

ship. For a given walk we assume that if he has not either reached the ship or fallen into the sea after, say, 10000 steps, he dies of thirst on the jetty.

To represent the jetty, we set up co-ordinates so that the *x*-axis runs along the middle of the jetty with the origin at the quay-end. *x* and *y* are measured in steps. The sailor starts his walk at the origin each time. The structure plan and script are as follows:

1. Initialize variables, including number of walks *n*
2. Repeat *n* simulated walks down the jetty:
 Start at the quay-end of the jetty
 While still on the jetty and still alive repeat:
 Get a random number *R* for the next step
 If *R* < 0.6 then
 Move forward (to the ship)
 Else if *R* < 0.8 then
 Move port (left)
 Else
 Move starboard
 If he got to the ship then
 Count that walk as a success
3. Compute and print estimated probability of reaching the ship
4. Stop.

```
% random walk
n = input( 'Number of walks: ' );
nsafe = 0;                    % number of times he makes it

for i = 1:n
  steps = 0;                  % each new walk ...
  x = 0;                      % ... starts at the origin
  y = 0;

  while x <= 50 & abs(y) <= 10 & steps < 1000
    steps = steps + 1;        % that's another step
    r = rand;                 % random number for that step
    if r < 0.6                % which way did he go?
      x = x + 1;              % maybe forward ...
    elseif r < 0.8
      y = y + 1;              % ... or to port ...
    else
      y = y - 1;              % ... or to starboard
    end;
  end;
```

```
      if x > 50
        nsafe = nsafe + 1;        % he actually made it this time!
      end;

    end;

    prob = 100 * nsafe / n;
    disp( prob );
```

A sample run of 100 walks gave an 93% probability of reaching the ship.

You can speed up the script by about 20% if you generate a vector of 1 000 random numbers, say, at the start of each walk (with `r = rand(1,1000);`) and then reference elements of the vector in the `while` loop, e.g.,

```
      if r(steps) < 0.6 ...
```

13.6 TRAFFIC FLOW

A major application of simulation is in modeling the traffic flow in large cities, in order to test different traffic light patterns before inflicting them on the real traffic. In this example we look at a very small part of the problem: how to simulate the flow of a single line of traffic through one set of traffic lights. We make the following assumptions (you can make additional or different ones if like):

1. Traffic travels straight, without turning.
2. The probability of a car arriving at the lights in a particular second is independent of what happened during the previous second. This is called a *Poisson process*. This probability (call it p) may be estimated by watching cars at the intersection and monitoring their arrival pattern. In this simulation we take $p = 0.3$.
3. When the lights are green, assume the cars move through at a steady rate of, say, eight every ten seconds.
4. In the simulation, we will take the basic time period to be ten seconds, so we want a display showing the length of the queue of traffic (if any) at the lights every ten seconds.
5. We will set the lights red or green for variable multiples of ten seconds.

The situation is modeled with a script file `traffic.m` which calls three function files: `go.m`, `stop.m`, and `prq.m`. Because the function files need access to a number of base workspace variables created by `traffic.m`, these variables are declared `global` in `traffic.m`, and in all three function files.

In this example the lights are red for 40 seconds (red = 4) and green for 20 seconds (green = 2). The simulation runs for 240 seconds (n = 24).

The script, traffic.m, is as follows:

```
clc
clear                   % clear out any previous garbage!
global CARS GTIMER GREEN LIGHTS RED RTIMER T

CARS = 0;               % number of cars in queue
GTIMER = 0;             % timer for green lights
GREEN = 2;              % period lights are green
LIGHTS = 'R';           % color of lights
n = 48;                 % number of 10-sec periods
p = 0.3;                % probability of a car arriving
RED = 4;                % period lights are red
RTIMER = 0;             % timer for red lights

for T = 1:n             % for each 10-sec period

  r = rand(1,10);       % 10 seconds means 10 random numbers
  CARS = CARS + sum(r < p);  % cars arriving in 10 seconds

  if LIGHTS == 'G'
    go                  % handles green lights
  else
    stop                % handles red lights
  end;

end;
```

The function files go.m, stop.m and prq.m (all separate M-files) are as follows:

```
% -------------------------------------------------------------
function go
global CARS GTIMER GREEN LIGHTS
GTIMER = GTIMER + 1;        % advance green timer
CARS = CARS - 8;            % let 8 cars through

if CARS < 0                 % ... there may have been < 8
  CARS = 0;
end;

prq;                        % display queue of cars
```

```
if GTIMER == GREEN          % check if lights need to change
   LIGHTS = 'R';
   GTIMER = 0;
end;

% -----------------------------------------------------------
function stop
global LIGHTS RED RTIMER
RTIMER = RTIMER + 1;        % advance red timer
prq;                        % display queue of cars

if RTIMER == RED            % check if lights must be changed
   LIGHTS = 'G';
   RTIMER = 0;
end;

% -----------------------------------------------------------
function prq
global CARS LIGHTS T
fprintf( '%3.0f ', T );     % display period number

if LIGHTS == 'R'            % display color of lights
   fprintf( 'R    ' );
else
   fprintf( 'G    ' );
end;

for i = 1:CARS              % display * for each car
   fprintf( '*' );
end;

fprintf( '\n' )             % new line
```

Typical output looks like this:

```
  1 R    ****
  2 R    ********
  3 R    ***********
  4 R    **************
  5 G    **********
  6 G    *****
  7 R    ********
  8 R    *************
  9 R    ****************
```

```
10 R   ********************
11 G   **************
12 G   **********
13 R   **************
14 R   *****************
15 R   ********************
16 R   ************************
17 G   **********************
18 G   ****************
19 R   *****************
20 R   *********************
21 R   ************************
22 R   ******************************
23 G   **************************
24 G   ***********************
```

From this particular run it seems that a traffic jam is building up, although more and longer runs are needed to see if this is really so. In that case, one can experiment with different periods for red and green lights in order to get an acceptable traffic pattern before setting the real lights to that cycle. Of course, we can get closer to reality by considering two-way traffic, and allowing cars to turn in both directions, and occasionally to break down, but this program gives the basic ideas.

13.7 NORMAL (GAUSSIAN) RANDOM NUMBERS

The function `randn` generates *Gaussian* or *normal* random numbers (as opposed to *uniform*) with a mean (μ) of 0 and a variance (σ^2) of 1.

- Generate 100 normal random numbers r with `randn(1,100)` and draw their histogram. Use the functions `mean(r)` and `std(r)` to find their mean and standard deviation (σ).
- Repeat with 1000 random numbers. The mean and standard deviation should be closer to 0 and 1 this time.

The functions `rand` and `randn` have separate generators, each with its own seed.

SUMMARY

- A simulation is a computer program written to mimic a "real-life" situation which is apparently based on chance.
- The pseudo-random number generator `rand` returns uniformly distributed random numbers in the range [0, 1), and is the basis of the simulations discussed in this chapter.

- `randn` generates normally distributed (Gaussian) random numbers.
- `rand('state', n)` enables the user to seed `rand` with any integer n. A seed may be obtained from `clock`, which returns the system clock time. `randn` may be seeded (independently) in a similar way.
- Each independent event being simulated requires one and only one random number.

EXERCISES

13.1 Write some statements to simulate spinning a coin 50 times using 0-1 vectors instead of a `for` loop. **Hints:** generate a vector of 50 random numbers, set up 0-1 vectors to represent the heads and tails, and use `double` and `char` to display them as a string of Hs and Ts.

13.2 In a game of Bingo the numbers 1 to 99 are drawn at random from a bag. Write a script to simulate the draw of the numbers (each number can be drawn only once), printing them ten to a line.

13.3 Generate some strings of 80 random alphabetic letters (lowercase only). For fun, see how many real words, if any, you can find in the strings.

13.4 A random number generator can be used to estimate π as follows (such a method is called a *Monte Carlo* method). Write a script which generates random points in a square with sides of length 2, say, and which counts what proportion of these points falls inside the circle of unit radius that fits exactly into the square. This proportion will be the ratio of the area of the circle to that of the square. Hence estimate π. (This is not a very efficient method, as you will see from the number of points required to get even a rough approximation.)

13.5 Write a script to simulate the progress of the short-sighted student in Chapter 16 (**Markov Processes**). Start him at a given intersection, and generate a random number to decide whether he moves toward the internet cafe or home, according to the probabilities in the transition matrix. For each simulated walk, record whether he ends up at home or in the cafe. Repeat a large number of times. The proportion of walks that end up in either place should approach the limiting probabilities computed using the Markov model described in Chapter 16. **Hint:** if the random number is less than 2/3 he moves toward the cafe (unless he is already at home or in the cafe, in which case that random walk ends), otherwise he moves toward home.

13.6 The aim of this exercise is to simulate bacteria growth.
Suppose that a certain type of bacteria divides or dies according to the following assumptions:

(a) during a fixed time interval, called a *generation*, a single bacterium divides into two identical replicas with probability p;

(b) if it does not divide during that interval, it dies;

(c) the offspring (called daughters) will divide or die during the next generation, independently of the past history (there may well be no offspring, in which case the colony becomes extinct).

Start with a single individual and write a script which simulates a number of generations. Take $p = 0.75$. The number of generations which you can simulate will depend on your computer system. Carry out a large number (e.g. 100) of such simulations. The probability of ultimate extinction, $p(E)$, may be estimated as the proportion of simulations that end in extinction. You can also estimate the mean size of the nth generation from a large number of simulations. Compare your estimate with the theoretical mean of $(2p)^n$.

Statistical theory shows that the expected value of the extinction probability $p(E)$ is the smaller of 1, and $(1 - p)/p$. So for $p = 0.75$, $p(E)$ is expected to be 1/3. But for $p \leq 0.5$, $p(E)$ is expected to be 1, which means that extinction is certain (a rather unexpected result). You can use your script to test this theory by running it for different values of p, and estimating $p(E)$ in each case.

13.7 Brian Hahn (the original author of this book) is indebted to a colleague, Gordon Kass, for suggesting this problem.

Dribblefire Jets Inc. make two types of aeroplane, the two-engined DFII, and the four-engined DFIV. The engines are terrible and fail with probability 0.5 on a standard flight (the engines fail independently of each other). The manufacturers claim that the planes can fly if at least half of their engines are working, i.e. the DFII will crash only if both its engines fail, while the DFIV will crash if all four, or if any three engines fail.

You have been commissioned by the Civil Aviation Board to ascertain which of the two models is less likely to crash. Since parachutes are expensive, the cheapest (and safest!) way to do this is to simulate a large number of flights of each model. For example, two calls of Math.random could represent one standard DFII flight: if both random numbers are less than 0.5, that flight crashes, otherwise it doesn't. Write a script which simulates a large number of flights of both models, and estimates the probability of a crash in each case. If you can run enough simulations, you may get a surprising result. (Incidentally, the probability of n engines failing on a given flight is given by the binomial distribution, but you do not need to use this fact in the simulation.)

13.8 Two players, A and B, play a game called *Eights*. They take it in turns to choose a number 1, 2 or 3, which may not be the same as the last number chosen (so if A starts with 2, B may only choose 1 or 3 at the next move). A starts, and may choose any of the three numbers for the first move. After each move, the number chosen is added to a common running total. If the total reaches 8 exactly, the player whose turn it

was wins the game. If a player causes the total to go over 8, the other player wins. For example, suppose A starts with 1 (total 1), B chooses 2 (total 3), A chooses 1 (total 4) and B chooses 2 (total 6). A would like to play 2 now, to win, but he can't because B cunningly played it on the last move, so A chooses 1 (total 7). This is even smarter, because B is forced to play 2 or 3, making the total go over 8 and thereby losing.

Write a script to simulate each player's chances of winning, if they always play at random.

13.9 If r is a normal random number with mean 0 and variance 1 (as generated by randn), it can be transformed into a random number X with mean μ and standard deviation σ by the relation

$$X = \sigma r + \mu.$$

In an experiment a Geiger counter is used to count the radio-active emissions of cobalt 60 over a 10-second period. After a large number of such readings are taken, the count rate is estimated to be normally distributed with a mean of 460 and a standard deviation of 20.

1. Simulate such an experiment 200 times by generating 200 random numbers with this mean and standard deviation. Plot the histogram (use 10 bins).

2. Repeat a few times to note how the histogram changes each time.

13.10 Radio-active carbon 11 has a decay-rate k of 0.0338 per minute, i.e. a particular C^{11} atom has a 3.38% chance of decaying in any one minute.

Suppose we start with 100 such atoms. We would like to simulate their fate over a period of 100 minutes, say. We want to end up with a bar graph showing how many atoms remain undecayed after 1, 2, ..., 100 minutes.

We need to simulate when each of the 100 atoms decays. This can be done, for each atom, by generating a random number r for each of the 100 minutes, until either $r > k$ (that atom decays), or the 100 minutes is up. If the atom decayed at time $t < 100$, increment the frequency distribution $f(t)$ by 1. $f(t)$ will be the number of atoms decaying at time t minutes.

Now convert the number $f(t)$ decaying each minute to the number $R(t)$ *remaining* each minute. If there are n atoms to start with, after one minute, the number $R(1)$ remaining will be $n - f(1)$, since $f(1)$ is the number decaying during the first minute. The number $R(2)$ remaining after two minutes will be $n - f(1) - f(2)$. In general, the number remaining after t minutes will be (in MATLAB notation)

```
R(t) = n - sum( f(1:t) )
```

FIGURE 13.1

Radio-active decay of carbon 11: simulated and theoretical.

Write a script to compute $R(t)$ and plot its bar graph. Superimpose on the bar graph the theoretical result, which is

$$R(t) = 100 \exp^{-kt}.$$

Typical results are shown in Fig. 13.1.

Introduction to Numerical Methods

THE OBJECTIVES OF THIS CHAPTER ARE TO INTRODUCE NUMERICAL METHODS FOR:

- Solving equations.
- Evaluating definite integrals.
- Solving systems of ordinary differential equations.
- Solving a parabolic partial differential equation.

CONTENTS

Equations 299
Newton's method ... 299
Complex roots *301*
The Bisection method 301
`fzero` 303
`roots` 303

Integration 304
The Trapezoidal rule 304
Simpson's rule 305
`quad` 306

Numerical differentiation .. 306
`diff` 307

First-order differential equations 308
Euler's method 308
Example: Bacteria growth 309
Alternative subscript notation .. 311
A predictor-corrector method ... 312

Linear ordinary differential equations (LODEs) 313

Runge-Kutta methods 313
A single differential equation 313
Systems of differential equations: Chaos ... 314

A major scientific use of computers is in finding numerical solutions to mathematical problems which have no analytical solutions (i.e., solutions which may be written down in terms of polynomials and standard mathematical functions). In this chapter we look briefly at some areas where *numerical methods* have been highly developed, e.g., solving non-linear equations, evaluating integrals, and solving differential equations.

14.1 EQUATIONS

In this section we consider how to solve equations in one unknown numerically. The usual way of expressing the problem is to say that we want to solve the equation $f(x) = 0$, i.e., we want to find its *root* (or roots). This process is also described as finding the *zeros* of $f(x)$. There is no general method for finding roots analytically for an arbitrary $f(x)$.

14.1.1 Newton's method

Newton's method is perhaps the easiest numerical method to implement for solving equations, and was introduced briefly in earlier chapters. It is an *iterative* method, meaning that it repeatedly attempts to improve an estimate of the root. If x_k is an approximation to the root, we can relate it to the next approximation x_{k+1} using the right-angle triangle in Fig. 14.1:

$$f'(x_k) = \frac{f(x_k) - 0}{x_k - x_{k+1}},$$

Essential MATLAB for Engineers and Scientists. https://doi.org/10.1016/B978-0-08-102997-8.00021-X

Passing additional parameters to an ODE solver............ 317

A partial differential equation.......... 318
Heat conduction 318

Complex variables and conformal mapping.......... 322
Joukowski airfoil.... 322

Other numerical methods.......... 323

Summary 325

Exercises 325

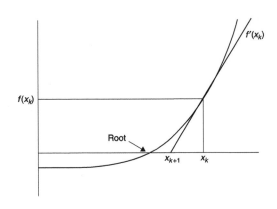

FIGURE 14.1
Newton's method.

where $f'(x)$ is df/dx. Solving for x_{k+1} gives

$$x_{k+1} = x_k - \frac{f(x_k)}{f'(x_k)}.$$

A structure plan to implement Newton's method is:

1. Input starting value x_0 and required relative error e
2. While relative error $|(x_k - x_{k-1})/x_k| \geq e$ repeat up to $k = 20$, say:
 $x_{k+1} = x_k - f(x_k)/f'(x_k)$
 Print x_{k+1} and $f(x_{k+1})$
3. Stop.

It is necessary to limit step 2 since the process may not converge.

A script using Newton's method (without the subscript notation) to solve the equation $x^3 + x - 3 = 0$ is given in Chapter 10. If you run it you will see that the values of x converge rapidly to the root.

As an exercise, try running the script with different starting values of x_0 to see whether the algorithm always converges.

If you have a sense of history, use Newton's method to find a root of $x^3 - 2x - 5 = 0$. This is the example used when the algorithm was first presented to the French Academy.

Also try finding a *non-zero* root of $2x = \tan(x)$, using Newton's method. You might have some trouble with this one. If you do, you will have discovered the one serious problem with Newton's method: it converges to a root only if the starting guess is 'close enough'. Since 'close enough' depends on the nature of $f(x)$ and on the root, one can obviously get into difficulties here. The only remedy is some intelligent trial-and-error work on the initial guess—this is

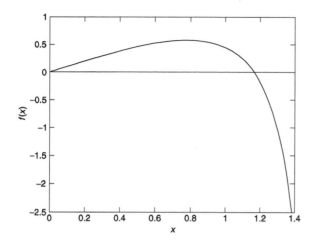

FIGURE 14.2
$f(x) = 2x - \tan(x)$.

made considerably easier by sketching $f(x)$ or plotting it with MATLAB (see Fig. 14.2).

If Newton's method fails to find a root, the Bisection method, discussed below, can be used.

14.1.1.1 Complex roots

Newton's method can also find complex roots, but only if the starting guess is complex. Use the script in Chapter 10 to find a complex root of $x^2 + x + 1 = 0$. Start with a complex value of 1 + i say, for x. Using this starting value for x gives the following output (if you replace disp([x f(x)]) in the script with disp(x)):

```
  0.0769 + 0.6154i
 -0.5156 + 0.6320i
 -0.4932 + 0.9090i
 -0.4997 + 0.8670i
 -0.5000 + 0.8660i
 -0.5000 + 0.8660i
Zero found
```

Since complex roots occur in complex conjugate pairs, the other root is $-0.5 - 0.866i$.

14.1.2 The Bisection method

Consider again the problem of solving the equation $f(x) = 0$, where

$$f(x) = x^3 + x - 3.$$

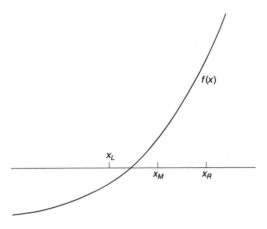

FIGURE 14.3
The Bisection method.

We attempt to find by inspection, or trial-and-error, two values of x, call them x_L and x_R, such that $f(x_L)$ and $f(x_R)$ have different signs, i.e., $f(x_L)f(x_R) < 0$. If we can find two such values, the root must lie somewhere in the interval between them, since $f(x)$ changes sign on this interval (see Fig. 14.3). In this example, $x_L = 1$ and $x_R = 2$ will do, since $f(1) = -1$ and $f(2) = 7$. In the Bisection method, we estimate the root by x_M, where x_M is the midpoint of the interval $[x_L, x_R]$, i.e.,

$$x_M = (x_L + x_R)/2. \tag{14.1}$$

Then if $f(x_M)$ has the same sign as $f(x_L)$, as drawn in the figure, the root clearly lies between x_M and x_R. We must then redefine the left-hand end of the interval as having the value of x_M, i.e., we let the new value of x_L be x_M. Otherwise, if $f(x_M)$ and $f(x_L)$ have *different* signs, we let the new value of x_R be x_M, since the root must lie between x_L and x_M in that case. Having redefined x_L or x_R, as the case may be, we bisect the new interval again according to Equation (14.1) and repeat the process until the distance between x_L and x_R is as small as we please.

The neat thing about this method is that, *before* starting, we can calculate how many bisections are needed to obtain a certain accuracy, given initial values of x_L and x_R. Suppose we start with $x_L = a$, and $x_R = b$. After the first bisection the worst possible error (E_1) in x_M is $E_1 = |a - b|/2$, since we are estimating the root as being at the midpoint of the interval $[a, b]$. The worst that can happen is that the root is actually at x_L or x_R, in which case the error is E_1. Carrying on like this, after n bisections the worst possible error E_n is given by $E_n = |a - b|/2^n$. If we want to be sure that this is less than some specified error E, we must see to it that n satisfies the inequality $|a - b|/2^n < E$, i.e.,

$$n > \frac{\log(|a - b|/E)}{\log(2)} \tag{14.2}$$

Since n is the number of bisections, it must be an integer. The smallest integer n that *exceeds* the right-hand side of Inequality (14.2) will do as the maximum number of bisections required to guarantee the given accuracy E.

The following scheme may be used to program the Bisection method. It will work for any function $f(x)$ that changes sign (in either direction) between the two values a and b, which must be found beforehand by the user.

1. Input a, b and E
2. Initialize x_L and x_R
3. Compute maximum bisections n from Inequality (14.2)
4. Repeat n times:
 Compute x_M according to Equation (14.1)
 If $f(x_L)f(x_M) > 0$ then
 Let $x_L = x_M$
 otherwise
 Let $x_R = x_M$
5. Display root x_M
6. Stop.

We have assumed that the procedure will not find the root exactly; the chances of this happening with real variables are infinitesimal.

The main advantage of the Bisection method is that it is guaranteed to find a root if you can find two starting values for x_L and x_R between which the function changes sign. You can also compute in advance the number of bisections needed to attain a given accuracy. Compared to Newton's method it is inefficient. Successive bisections do not necessarily move closer to the root, as usually happens with Newton's method. In fact, it is interesting to compare the two methods on the same function to see how many more steps the Bisection method requires than Newton's method. For example, to solve the equation $x^3 + x - 3 = 0$, the Bisection method takes 21 steps to reach the same accuracy as Newton's in five steps.

14.1.3 fzero

The MATLAB function `fzero(@f, a)` finds the zero nearest to the value `a` of the function `f` represented by the function `f.m`.

Use it to find a zero of $x^3 + x - 3$.

`fzero` doesn't appear to be able to find complex roots.

14.1.4 roots

The MATLAB function M-file `roots(c)` finds all the roots (zeros) of the polynomial with coefficients in the vector `c`. See `help` for details.

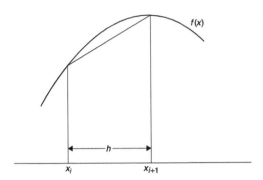

FIGURE 14.4
The Trapezoidal rule.

Use it to find a zero of $x^3 + x - 3$.

14.2 INTEGRATION

Although most 'respectable' mathematical functions can be differentiated analytically, the same cannot be said for integration. There are no general rules for integrating, as there are for differentiating. For example, the indefinite integral of a function as simple as e^{-x^2} cannot be found analytically. We therefore need numerical methods for evaluating integrals.

This is actually quite easy, and depends on the fact that the definite integral of a function $f(x)$ between the limits $x = a$ and $x = b$ is equal to the area under $f(x)$ bounded by the x-axis and the two vertical lines $x = a$ and $x = b$. So all numerical methods for integrating simply involve more or less ingenious ways of estimating the area under $f(x)$.

14.2.1 The Trapezoidal rule

The Trapezoidal (or Trapezium) rule is fairly simple to program. The area under $f(x)$ is divided into vertical panels each of width h, called the *step-length*. If there are n such panels, then $nh = b - a$, i.e., $n = (b - a)/h$. If we join the points where successive panels cut $f(x)$, we can estimate the area under $f(x)$ as the sum of the area of the resulting trapezia (see Fig. 14.4). If we call this approximation to the integral S, then

$$S = \frac{h}{2}\left[f(a) + f(b) + 2\sum_{i=1}^{n-1} f(x_i)\right], \tag{14.3}$$

where $x_i = a + ih$. Equation (14.3) is the Trapezoidal rule, and provides an estimate for the integral

$$\int_a^b f(x)\,dx.$$

Here is a function to implement the Trapezoidal rule:

```
function y = trap( fn, a, b, h )
n = (b-a)/h;
x = a + [1:n-1]*h;
y = sum(feval(fn, x));
y = h/2*(feval(fn, a) + feval(fn, b) + 2*y);
```

Note:

1. Since the summation in the rule is implemented with a vectorized formula rather than a `for` loop (to save time), the function to be integrated must use array operators where appropriate in its M-file implementation.
2. The user must choose h in such a way that the number of steps n will be an integer—a check for this could be built in.

As an exercise, integrate $f(x) = x^3$ between the limits 0 and 4 (remember to write `x.^3` in the function M-file). Call `trap` as follows:

```
s = trap(@f, 0, 4, h);
```

With $h = 0.1$, the estimate is 64.04, and with $h = 0.01$ it is 64.0004 (the exact integral is 64). You will find that as h gets smaller, the estimate gets more accurate.

This example assumes that $f(x)$ is a continuous function which may be evaluated at any x. In practice, the function could be defined at discrete points supplied as results of an experiment. For example, the speed of an object $v(t)$ might be measured every so many seconds, and one might want to estimate the distance traveled as the area under the speed-time graph. In this case, `trap` would have to be changed by replacing `fn` with a vector of function values. This is left as an exercise for the curious. Alternatively, you can use the MATLAB function `interp1` to interpolate the data. See `help`.

14.2.2 Simpson's rule

Simpson's rule is a method of numerical integration which is a good deal more accurate than the Trapezoidal rule, and should always be used before you try

anything fancier. It also divides the area under the function to be integrated, $f(x)$, into vertical strips, but instead of joining the points $f(x_i)$ with straight lines, every set of three such successive points is fitted with a parabola. To ensure that there are always an even number of panels, the step-length h is usually chosen so that there are $2n$ panels, i.e., $n = (b - a)/(2h)$.

Using the same notation as above, Simpson's rule estimates the integral as

$$S = \frac{h}{3}\left[f(a) + f(b) + 2\sum_{i=1}^{n-1} f(x_{2i}) + 4\sum_{i=1}^{n} f(x_{2i-1}) \right].$$ (14.4)

Coding this formula into a function M-file is left as an exercise.

If you try Simpson's rule on $f(x) = x^3$ between any limits, you will find rather surprisingly, that it gives the same result as the exact mathematical solution. This is a nice extra benefit of the rule: it integrates cubic polynomials exactly (which can be proved).

14.2.3 quad

Not surprisingly, MATLAB has a function `quad` to carry out numerical integration, or *quadrature* as it is also called. See `help`.

You may think there is no point in developing our own function files to handle these numerical procedures when MATLAB has its own. If you have got this far, you should be curious enough to want to know how they work, rather than treating them simply as 'black boxes'.

14.3 NUMERICAL DIFFERENTIATION

The *Newton quotient* for a function $f(x)$ is given by

$$\frac{f(x+h) - f(x)}{h},$$ (14.5)

where h is 'small'. As h tends to zero, this quotient approaches the first derivative, df/dx. The Newton quotient may therefore be used to estimate a derivative numerically. It is a useful exercise to do this with a few functions for which you know the derivatives. This way you can see how small you can make h before rounding errors cause problems. Such errors arise because expression (14.5) involves subtracting two terms that eventually become equal when the limit of the computer's accuracy is reached.

As an example, the following script uses the Newton quotient to estimate $f'(x)$ for $f(x) = x^2$ (which must be supplied as a function file `f.m`) at $x = 2$, for smaller and smaller values of h (the exact answer is 4).

```
h = 1;
x = 2;
format short e
for i = 1:20
  nq = (f(x+h) - f(x))/h;
  disp( [h nq] )
  h = h / 10;
end
```

Output:

```
1                5
1.0000e-001   4.1000e+000
1.0000e-002   4.0100e+000
1.0000e-003   4.0010e+000
1.0000e-004   4.0001e+000
1.0000e-005   4.0000e+000
1.0000e-006   4.0000e+000
1.0000e-007   4.0000e+000
1.0000e-008   4.0000e+000
1.0000e-009   4.0000e+000
1.0000e-010   4.0000e+000
1.0000e-011   4.0000e+000
1.0000e-012   4.0004e+000
1.0000e-013   3.9968e+000
1.0000e-014   4.0856e+000
1.0000e-015   3.5527e+000
1.0000e-016            0
...
```

The results show that the best h for this particular problem is about 10^{-8}. But for h much smaller than this the estimate becomes totally unreliable.

Generally, the best h for a given problem can only be found by trial and error. Finding it can be a non-trivial exercise. This problem does not arise with numerical integration, because numbers are *added* to find the area, not subtracted.

14.3.1 diff

If x is a row or column vector

```
[x(1)  x(2)  ...  x(n)]
```

then the MATLAB function diff(x) returns a vector of differences between adjacent elements:

```
[x(2)-x(1)  x(3)-x(2)  ...  x(n)-x(n-1)]
```

The output vector is one element shorter than the input vector.

In certain problems, `diff` is helpful in finding approximate derivatives, e.g., if `x` contains displacements of an object every `h` seconds, `diff(x)/h` will be its speed.

14.4 FIRST-ORDER DIFFERENTIAL EQUATIONS

The most interesting situations in real life that we may want to model, or represent quantitatively, are usually those in which the variables change in time (e.g., biological, electrical or mechanical systems). If the changes are continuous, the system can often be represented with equations involving the derivatives of the dependent variables. Such equations are called *differential equations*. The main aim of a lot of modeling is to be able to write down a set of differential equations (DEs) that describe the system being studied as accurately as possible. Very few DEs can be solved analytically, so once again, numerical methods are required. We will consider the simplest method of numerical solution in this section: Euler's method (Euler rhymes with 'boiler'). We also consider briefly how to improve it.

14.4.1 Euler's method

In general we want to solve a first-order DE (strictly an ordinary—ODE) of the form

$$\mathrm{d}y/\mathrm{d}x = f(x, y), \quad y(0) \text{ given.}$$

Euler's method for solving this DE numerically consists of replacing $\mathrm{d}y/\mathrm{d}x$ with its Newton quotient, so that the DE becomes

$$\frac{y(x + h) - y(x)}{h} = f(x, y).$$

After a slight rearrangement of terms, we get

$$y(x + h) = y(x) + hf(x, y). \tag{14.6}$$

Solving a DE numerically is such an important and common problem in science and engineering that it is worth introducing some general notation at this point. Suppose we want to integrate the DE over the interval $x = a$ ($a = 0$ usually) to $x = b$. We break this interval up into m steps of length h, so

$$m = (b - a)/h$$

(this is the same as the notation used in the update process of Chapter 11, except that dt used there has been replaced by the more general h here).

For consistency with MATLAB's subscript notation, if we define y_i as $y(x_i)$ (the Euler estimate at the *beginning* of step i), where $x_i = (i - 1)h$, then $y_{i+1} = y(x + h)$, at the end of step i. We can then replace Equation (14.6) by the iterative scheme

$$y_{i+1} = y_i + hf(x_i, y_i), \tag{14.7}$$

where $y_1 = y(0)$. Recall from Chapter 11 that this notation enables us to generate a vector y which we can then plot. Note also the striking similarity between Equation (14.7) and the equation in Chapter 11 representing an update process. This similarity is no coincidence. Update processes are typically modeled by DEs, and Euler's method provides an approximate solution for such DEs.

14.4.2 Example: Bacteria growth

Suppose a colony of 1000 bacteria is multiplying at the rate of $r = 0.8$ per hour per individual (i.e. an individual produces an average of 0.8 offspring every hour). How many bacteria are there after 10 hours? Assuming that the colony grows continuously and without restriction, we can model this growth with the DE

$$dN/dt = rN, \quad N(0) = 1000, \tag{14.8}$$

where $N(t)$ is the population size at time t. This process is called *exponential growth*. Equation (14.8) may be solved analytically to give the well-known formula for exponential growth:

$$N(t) = N(0)e^{rt}.$$

To solve Equation (14.8) numerically, we apply Euler's algorithm to it to get

$$N_{i+1} = N_i + rhN_i, \tag{14.9}$$

where the initial value $N_1 = 1000$.

It is very easy to program Euler's method. The following script implements Equation (14.9), taking $h = 0.5$. It also computes the exact solution for comparison.

```
h = 0.5;
r = 0.8;
a = 0;
b = 10;
m = (b - a) / h;
N = zeros(1, m+1);
```

Time (hours)	Euler	Predictor-Corrector	Exact
0.0	1000	1000	1000
0.5	1400	1480	1492
1.0	1960	2190	2226
1.5	2744	3242	3320
2.0	3842	4798	4953
...			
5.0	28925	50422	54598
...			
8.0	217795	529892	601845
...			
10.0	836683	2542344	2980958

Table 14.1 Bacteria growth.

```
N(1) = 1000;
t = a:h:b;

for i = 1:m
  N(i+1) = N(i) + r * h * N(i);
end

Nex = N(1) * exp(r * t);
format bank
disp( [t' N' Nex'] )

plot(t, N ), xlabel( 'Hours' ), ylabel( 'Bacteria' )
hold on
plot(t, Nex ), hold off
```

Results are shown in Table 14.1, and also in Fig. 14.5. The Euler solution is not too good. In fact, the error gets worse at each step, and after 10 hours of bacteria time it is about 72 per cent. The numerical solution will improve if we make h smaller, but there will always be some value of t where the error exceeds some acceptable limit.

In some cases, Euler's method performs better than it does here, but there are other numerical methods which always do better than Euler. Two of them are discussed below. More sophisticated methods may be found in most textbooks on numerical analysis. However, Euler's method may always be used as a first approximation as long as you realize that errors may arise.

FIGURE 14.5
Bacteria growth: (a) Euler's method; (b) the exact solution.

14.4.3 Alternative subscript notation

Equation (14.9) is an example of a *finite difference scheme*. The conventional finite difference notation is for the initial value to be represented by N_0, i.e. with subscript $i = 0$. N_i is then the estimate at the *end* of step i. If you want the MATLAB subscripts in the Euler solution to be the same as the finite difference subscripts, the initial value N_0 must be represented by the MATLAB scalar N0, and you have to compute N(1) separately, before the for loop starts. You also have to display or plot the initial values separately since they will no longer be included in the MATLAB vectors t, N and Nex (which now have m instead of $m + 1$ elements). Here is a complete script to generate the Euler solution using finite difference subscripts:

```
h = 0.5;
r = 0.8;
a = 0;
b = 10;
m = (b - a) / h;
N = zeros(1, m);      % one less element now
N0 = 1000;
N(1) = N0 + r*h*N0;   % no longer 'self-starting'

for i = 2:m
  N(i) = N(i-1) + r * h * N(i-1); %finite difference notation
end

t = a+h:h:b;          % exclude initial time = a
```

```
Nex = N0 * exp(r * t);
disp( [a N0 N0] )    % display initial values separately
disp( [t' N' Nex'] )

plot(a, N0)          % plot initial values separately
hold on
plot(t, N ), xlabel( 'Hours' ), ylabel( 'Bacteria' )
plot(t, Nex ), hold off
```

14.4.4 A predictor-corrector method

One improvement on the numerical solution of the first-order DE

$$\mathrm{d}y/\mathrm{d}x = f(x, y), \quad y(0) \text{ given},$$

is as follows. The Euler approximation, which we are going to denote by an asterisk, is given by

$$y_{i+1}^* = y_i + h f(x_i, y_i) \tag{14.10}$$

But this formula favors the old value of y in computing $f(x_i, y_i)$ on the right-hand side. Surely it would be better to say

$$y_{i+1}^* = y_i + h[f(x_{i+1}, y_{i+1}^*) + f(x_i, y_i)]/2, \tag{14.11}$$

where $x_{i+1} = x_i + h$, since this also involves the new value y_{i+1}^* in computing f on the right-hand side? The problem of course is that y_{i+1}^* is as yet unknown, so we can't use it on the right-hand side of Equation (14.11). But we could use Euler to estimate (predict) y_{i+1}^* from Equation (14.10) and then use Equation (14.11) to correct the prediction by computing a *better* version of y_{i+1}^*, which we will call y_{i+1}. So the full procedure is:

> Repeat as many times as required:
> Use Euler to predict: $y_{i+1}^* = y_i + h f(x_i, y_i)$
> Then correct y_{i+1}^* to: $y_{i+1} = y_i + h[f(x_{i+1}, y_{i+1}^*) + f(x_i, y_i)]/2$.

This is called a *predictor-corrector* method. The script above can easily be adapted to this problem. The relevant lines of code are:

```
for i = 1:m     % m steps of length dt
  ne(i+1) = ne(i) + r * h * ne(i);
  np = nc(i) + r * h * nc(i);
  nc(i+1) = nc(i) + r * h * (np + nc(i))/2;
  disp( [t(i+1) ne(i+1) nc(i+1) nex(i+1)] )
end;
```

ne stands for the 'straight' (uncorrected) Euler solution, np is the Euler predictor (since this is an intermediate result a vector is not needed for np), and nc is the corrector. The worst error is now only 15 per cent. This is much better than the uncorrected Euler solution, although there is still room for improvement.

14.5 LINEAR ORDINARY DIFFERENTIAL EQUATIONS (LODES)

Linear ODEs with constant coefficients may be solved analytically in terms of *matrix exponentials*, which are represented in MATLAB by the function expm. For an example see **MATLAB Help: Mathematics: Matrices and Linear Algebra: Matrix Powers and Exponentials**.

14.6 RUNGE-KUTTA METHODS

There are a variety of algorithms, under the general name of Runge-Kutta, which can be used to integrate systems of ODEs. The formulae involved are rather complicated; they can be found in most books on numerical analysis.

However, as you may have guessed, MATLAB has plenty of ODE solvers, which are discussed in **MATLAB Help: Mathematics: Differential Equations**. Among them are ode23 (second/third order) and ode45 (fourth/fifth order), which implement Runge-Kutta methods. (The *order* of a numerical method is the power of h (i.e., dt) in the leading error term. Since h is generally very small, the higher the power, the smaller the error). We will demonstrate the use of ode23 and ode45 here, first with a single first-order DE, and then with systems of such equations.

14.6.1 A single differential equation

Here's how to use ode23 to solve the bacteria growth problem, Equation (14.8):

$$dN/dt = rN, \quad N(0) = 1000.$$

1. Start by writing a function file for the *right-hand side* of the DE to be solved. The function must have input variables t and N in this case (i.e., independent and dependent variables of the DE), *in that order*, e.g., create the function file **f.m** as follows:

   ```
   function y = f(t, Nr)
   y = 0.8 * Nr;
   ```

2. Now enter the following statements in the Command Window:

   ```
   a = 0;
   b = 10;
   ```

```
n0 = 1000;
[t, Nr] = ode23(@f, [a:0.5:b], n0);
```

3. Note the input arguments of `ode23`:
 `@f`: a handle for the function `f`, which contains the right-hand side of the DE;
 `[a:0.5:b]`: a vector (`tspan`) specifying the range of integration. If `tspan` has two elements (`[a b]`) the solver returns the solution evaluated at every integration step (the solver chooses the integration steps and may vary them). This form would be suitable for plotting. However, if you want to display the solution at regular time intervals, as we want to here, use the form of `tspan` with three elements as above. The solution is then returned evaluated at each time in `tspan`. The accuracy of the solution is not affected by the form of `tspan` used.
 `n0`: the initial value of the solution N.

4. The output arguments are two vectors: the solutions `Nr` at times `t`. For 10 hours `ode23` gives a value of 2961338 bacteria. From the exact solution in Table 14.1 we see that the error here is only 0.7 per cent.

If the solutions you get from `ode23` are not accurate enough, you can request greater accuracy with an additional optional argument. See `help`.

If you need still more accurate numerical solutions, you can use `ode45` instead. It gives a final value for the bacteria of 2981290—an error of about 0.01%.

14.6.2 Systems of differential equations: Chaos

The reason that weather prediction is so difficult and forecasts are so erratic is no longer thought to be the complexity of the system but the nature of the DEs modeling it. These DEs belong to a class referred to as *chaotic*. Such equations will produce wildly different results when their initial conditions are changed infinitesimally. In other words, accurate weather prediction depends crucially on the accuracy of the measurements of the initial conditions.

Edward Lorenz, a research meteorologist, discovered this phenomenon in 1961. Although his original equations are far too complex to consider here, the following much simpler system has the same essential chaotic features:

$$dx/dt = 10(y - x), \tag{14.12}$$

$$dy/dt = -xz + 28x - y, \tag{14.13}$$

$$dz/dt = xy - 8z/3. \tag{14.14}$$

This system of DEs may be solved very easily with the MATLAB ODE solvers. The idea is to solve the DEs with certain initial conditions, plot the solution, then change the initial conditions very slightly, and superimpose the new solution over the old one to see how much it has changed.

We begin by solving the system with the initial conditions $x(0) = -2$, $y(0) = -3.5$ and $z(0) = 21$.

1. Write a function file `lorenz.m` to represent the right-hand sides of the system as follows:

```
function f = lorenz(t, x)
f = zeros(3,1);
f(1) = 10 * (x(2) - x(1));
f(2) = -x(1) * x(3) + 28 * x(1) - x(2);
f(3) = x(1) * x(2) - 8 * x(3) / 3;
```

The three elements of the MATLAB vector x, i.e., `x(1)`, `x(2)` and `x(3)`, represent the three dependent scalar variables x, y and z respectively. The elements of the vector `f` represent the right-hand sides of the three DEs. When a vector is returned by such a DE function it must be a column vector, hence the statement

```
f = zeros(3,1);
```

2. Now use the following commands to solve the system from $t = 0$ to $t = 10$, say:

```
x0 = [-2 -3.5 21];    % initial values in a vector
[t, x] = ode45(@lorenz, [0 10], x0);
plot(t,x)
```

Note that we are use `ode45` now, since it is more accurate.
You will see three graphs, for x, y and z (in different colors).

3. It's easier to see the effect of changing the initial values if there is only one graph in the figure to start with. It is in fact best to plot the solution $y(t)$ on its own.
The MATLAB solution x is actually a matrix with three columns (as you can see from `whos`). The solution $y(t)$ that we want will be the second column, so to plot it by itself use the command

```
plot(t,x(:,2),'g')
```

Then keep the graph on the axes with the command `hold`.

Now we can see the effect of changing the initial values. Let's just change the initial value of $x(0)$, from -2 to -2.04—that's a change of only 2 per cent, and in only one of the three initial values. The following commands will do this, solve the DEs, and plot the new graph of $y(t)$ (in a different color):

```
x0 = [-2.04 -3.5 21];
[t, x] = ode45(@lorenz, [0 10], x0);
plot(t,x(:,2),'r')
```

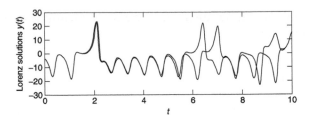

FIGURE 14.6
Chaos?

You should see (Fig. 14.6) that the two graphs are practically indistinguishable until t is about 1.5. The discrepancy grows quite gradually, until t reaches about 6, when the solutions suddenly and shockingly flip over in opposite directions. As t increases further, the new solution bears no resemblance to the old one.

Now solve the system (14.12)–(14.14) with the original initial values using ode23 this time:

```
x0 = [-2 -3.5 21];
[t,x] = ode23(@lorenz, [0 10], x0);
```

Plot the graph of $y(t)$ only—x(:,2)—and then superimpose the ode45 solution *with the same initial values* (in a different color).

A strange thing happens—the solutions begin to deviate wildly for $t > 1.5$! The initial conditions are the same—the only difference is the order of the Runge-Kutta method.

Finally solve the system with ode23s and superimpose the solution. (The s stands for 'stiff'. For a stiff DE, solutions can change on a time scale that is very short compared to the interval of integration.) The ode45 and ode23s solutions only start to diverge at $t > 5$.

The explanation is that ode23, ode23s and ode45 all have numerical inaccuracies (if one could compare them with the exact solution—which incidentally can't be found). However, the numerical inaccuracies are *different* in the three cases. This difference has the same effect as starting the numerical solution with very slightly different initial values.

How do we ever know when we have the 'right' numerical solution? Well, we don't—the best we can do is increase the accuracy of the numerical method until no further wild changes occur over the interval of interest. So in our example we can only be pretty sure of the solution for $t < 5$ (using ode23s or ode45). If that's not good enough, you have to find a more accurate DE solver.

So beware: 'chaotic' DEs are very tricky to solve!

Incidentally, if you want to see the famous 'butterfly' picture of chaos, just plot x against z as time increases (the resulting graph is called a *phase plane* plot). The following command will do the trick:

```
plot(x(:,1), x(:,3))
```

What you will see is a static 2-D projection of the *trajectory*, i.e., the solution developing in time. **Demos** in the MATLAB Launch Pad include an example which enables you to see the trajectory evolving dynamically in 3-D (**Demos: Graphics: Lorenz attractor animation**).

14.6.3 Passing additional parameters to an ODE solver

In the above examples of the MATLAB ODE solvers the coefficients in the right-hand sides of the DEs (e.g., the value 28 in Equation (14.13)) have all been constants. In a real modeling situation, you will most likely want to change such coefficients frequently. To avoid having to edit the function files each time you want to change a coefficient, you can pass the coefficients as additional parameters to the ODE solver, which in turn passes them to the DE function. To see how this may be done, consider the Lotka-Volterra *predator-prey* model:

$$dx/dt = px - qxy \qquad (14.15)$$
$$dy/dt = rxy - sy, \qquad (14.16)$$

where $x(t)$ and $y(t)$ are the prey and predator population sizes at time t, and p, q, r and s are biologically determined parameters. For this example, we take $p = 0.4$, $q = 0.04$, $r = 0.02$, $s = 2$, $x(0) = 105$ and $y(0) = 8$.

First, write a function M-file, `volterra.m` as follows:

```
function f = volterra(t, x, p, q, r, s)
f = zeros(2,1);
f(1) = p*x(1) - q*x(1)*x(2);
f(2) = r*x(1)*x(2) - s*x(2);
```

Then enter the following statements in the Command Window, which generate the characteristically oscillating graphs in Fig. 14.7:

```
p = 0.4; q = 0.04; r = 0.02; s = 2;
[t,x] = ode23(@volterra,[0 10],[105; 8],[],p,q,r,s);
plot(t, x)
```

Note:

- The additional parameters (p, q, r and s) have to follow the fourth input argument (options—see help) of the ODE solver. If no options have been set (as in our case), use [] as a placeholder for the options parameter.

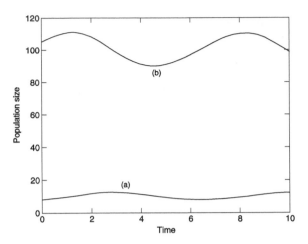

FIGURE 14.7
Lotka-Volterra model: (a) predator; (b) prey.

You can now change the coefficients from the Command Window and get a new solution, without editing the function file.

14.7 A PARTIAL DIFFERENTIAL EQUATION

The numerical solution of partial differential equations (PDEs) is a vast subject, which is beyond the scope of this book. However, a class of PDEs called *parabolic* often lead to solutions in terms of sparse matrices, which were mentioned briefly in Chapter 16. One such example is considered in this section.

14.7.1 Heat conduction

The conduction of heat along a thin uniform rod may be modeled by the partial differential equation

$$\frac{\partial u}{\partial t} = \frac{\partial^2 u}{\partial x^2}, \tag{14.17}$$

where $u(x, t)$ is the temperature distribution a distance x from one end of the rod at time t, and assuming that no heat is lost from the rod along its length.

Half the battle in solving PDEs is mastering the notation. We set up a rectangular grid, with step-lengths of h and k in the x and t directions respectively. A general point on the grid has co-ordinates $x_i = ih$, $y_j = jk$. A concise notation for $u(x, t)$ at x_i, y_j is then simply $u_{i,j}$.

Truncated Taylor series may then be used to approximate the PDE by a *finite difference scheme*. The left-hand side of Equation (14.17) is usually approximated

by a *forward difference*:

$$\frac{\partial u}{\partial t} = \frac{u_{i,j+1} - u_{i,j}}{k}$$

One way of approximating the right-hand side of Equation (14.17) is by the scheme

$$\frac{\partial^2 u}{\partial x^2} = \frac{u_{i+1,j} - 2u_{i,j} + u_{i-1,j}}{h^2}. \tag{14.18}$$

This leads to a scheme, which although easy to compute, is only conditionally stable.

If however we replace the right-hand side of the scheme in Equation (14.18) by the mean of the finite difference approximation on the jth and $(j+1)$th time rows, we get (after a certain amount of algebra!) the following scheme for Equation (14.17):

$$-ru_{i-1,j+1} + (2+2r)u_{i,j+1} - ru_{i+1,j+1} = ru_{i-1,j} + (2-2r)u_{i,j} + ru_{i+1,j}, \tag{14.19}$$

where $r = k/h^2$. This is known as the Crank-Nicolson *implicit* method, since it involves the solution of a system of simultaneous equations, as we shall see.

To illustrate the method numerically, let's suppose that the rod has a length of 1 unit, and that its ends are in contact with blocks of ice, i.e., the *boundary conditions* are

$$u(0,t) = u(1,t) = 0. \tag{14.20}$$

Suppose also that the initial temperature (*initial condition*) is

$$u(x,0) = \begin{cases} 2x, & 0 \le x \le 1/2, \\ 2(1-x), & 1/2 \le x \le 1. \end{cases} \tag{14.21}$$

(this situation could come about by heating the center of the rod for a long time, with the ends kept in contact with the ice, removing the heat source at time $t = 0$.) This particular problem has symmetry about the line $x = 1/2$; we exploit this now in finding the solution.

If we take $h = 0.1$ and $k = 0.01$, we will have $r = 1$, and Equation (14.19) becomes

$$-u_{i-1,j+1} + 4u_{i,j+1} - u_{i+1,j+1} = u_{i-1,j} + u_{i+1,j}. \tag{14.22}$$

Putting $j = 0$ in Equation (14.22) generates the following set of equations for the unknowns $u_{i,1}$ (i.e., after one time step k) up to the midpoint of the rod,

which is represented by $i = 5$, i.e., $x = ih = 0.5$. The subscript $j = 1$ has been dropped for clarity:

$$0 + 4u_1 - u_2 = 0 + 0.4$$
$$-u_1 + 4u_2 - u_3 = 0.2 + 0.6$$
$$-u_2 + 4u_3 - u_4 = 0.4 + 0.8$$
$$-u_3 + 4u_4 - u_5 = 0.6 + 1.0$$
$$-u_4 + 4u_5 - u_6 = 0.8 + 0.8.$$

Symmetry then allows us to replace u_6 in the last equation by u_4. These equations can be written in matrix form as

$$
\begin{bmatrix}
4 & -1 & 0 & 0 & 0 \\
-1 & 4 & -1 & 0 & 0 \\
0 & -1 & 4 & -1 & 0 \\
0 & 0 & -1 & 4 & -1 \\
0 & 0 & 0 & -2 & 4
\end{bmatrix}
\begin{bmatrix}
u_1 \\
u_2 \\
u_3 \\
u_4 \\
u_5
\end{bmatrix}
=
\begin{bmatrix}
0.4 \\
0.8 \\
1.2 \\
1.6 \\
1.6
\end{bmatrix}. \tag{14.23}
$$

The matrix (**A**) on the left of Equations (14.23) is known as a *tridiagonal* matrix. Having solved for the $u_{i,1}$ we can then put $j = 1$ in Equation (14.22) and proceed to solve for the $u_{i,2}$, and so on. The system (14.23) can of course be solved directly in MATLAB with the left division operator. In the script below, the general form of Equations (14.23) is taken as

$$\mathbf{Av} = \mathbf{g}. \tag{14.24}$$

Care needs to be taken when constructing the matrix **A**. The following notation is often used:

$$
\mathbf{A} =
\begin{bmatrix}
b_1 & c_1 & & & & \\
a_2 & b_2 & c_2 & & & \\
& a_3 & b_3 & c_3 & & \\
& & & \ddots & & \\
& & & a_{n-1} & b_{n-1} & c_{n-1} \\
& & & & a_n & b_n
\end{bmatrix}.
$$

A is an example of a sparse matrix (see Chapter 16).

The script below implements the general Crank-Nicolson scheme of Equation (14.19) to solve this particular problem over 10 time steps of $k = 0.01$. The step-length is specified by $h = 1/(2n)$ because of symmetry. r is therefore not restricted to the value 1, although it takes this value here. The script exploits the sparsity of **A** by using the `sparse` function.

```
format compact
n = 5;
k = 0.01;
h =  1 / (2 * n);                        % symmetry assumed
r = k / h ^ 2;

% set up the (sparse) matrix A
b = sparse(1:n, 1:n, 2+2*r, n, n);       % b(1) .. b(n)
c = sparse(1:n-1, 2:n, -r, n, n);        % c(1) .. c(n-1)
a = sparse(2:n, 1:n-1, -r, n, n);        % a(2) ..
A = a + b + c;
A(n, n-1) = -2 * r;                      % symmetry: a(n)
full(A)                                  %
disp(' ')

u0 = 0;                  % boundary condition (Eq 19.20)
u = 2*h*[1:n]            % initial conditions (Eq 19.21)
u(n+1) = u(n-1);         % symmetry
disp([0 u(1:n)])

for t = k*[1:10]
  g = r * ([u0 u(1:n-1)] + u(2:n+1)) ...
                      + (2 - 2 * r) * u(1:n);  % Eq 19.19
  v = A\g';                  % Eq 19.24
  disp([t v'])
  u(1:n) = v;
  u(n+1) = u(n-1);          % symmetry
end
```

Note:

- to preserve consistency between the formal subscripts of Equation (14.19) etc. and MATLAB subscripts, u_0 (the boundary value) is represented by the scalar u0.

In the following output the first column is time, and subsequent columns are the solutions at intervals of h along the rod:

```
      0    0.2000    0.4000    0.6000    0.8000    1.0000
 0.0100    0.1989    0.3956    0.5834    0.7381    0.7691
 0.0200    0.1936    0.3789    0.5397    0.6461    0.6921
   ...
 0.1000    0.0948    0.1803    0.2482    0.2918    0.3069
```

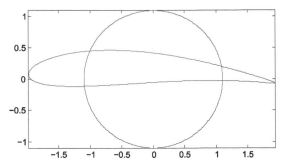

FIGURE 14.8
Map of circle to an airfoil: illustration of the application of the Joukowski transformation in the complex plane.

MATLAB has some built-in PDE solvers. See **MATLAB Help: Mathematics: Differential Equations: Partial Differential Equations**.

14.8 COMPLEX VARIABLES AND CONFORMAL MAPPING

In this section one application of the complex-variable capabilities in MATLAB is demonstrated. It is the transformation of a circle to a Joukowski airfoil.

Joukowski airfoil

The solution of the flow around a circular cylinder with circulation in a cross flow can be used to predict the flow around thin airfoils. We can transform the local geometry of the cylinder into an ellipse, an airfoil or a flat plate without influencing the geometry if the far field. This procedure is known as conformal mapping. If we interpret the Cartesian coordinates as the coordinates the plane of complex numbers $z = x + iy$, where x and y are real numbers, $i = \sqrt{-1}$, x is the real part of z and y is the imaginary part of z, then doing this allows us to use complex variable theory to solve two-dimensional potential flow problems. We are *not* going to examine complex variable theory here. Instead, we will give an example of the application of one of the ideas to illustrate that the circle can be transformed into an airfoil. Table 14.2 is a MATLAB script that does this. MATLAB is very useful for this problem because it does complex arithmetic. The steps are outlined in the table. The result of executing this code is illustrated in Fig. 14.8. The figure shows the circle and the airfoil that is mapped from it. Each point on the circle corresponds to a unique point on the airfoil. The potential at each point on the cylinder is the same as the corresponding point on the airfoil. This is how the solution of the circle problem is mapped to solve the flow around the airfoil. What is different is the distance between points and, hence, the velocity and pressure distributions on the airfoil must be determined from the mapped distribution of the potential. The far field is not affected by the transformation.

Table 14.2 MATLAB file used to produce Fig. 14.8.

```
% Joukowski transformation MATLAB code
%
% Example of conformal mapping of a circle to an airfoil:
% A problem in the field of aerodynamics
% Daniel T. Valentine ...................... 2009/2018.
% Circle in (xp,yp) plane: R = sqrt(xp^2 + yp^2), R > 1
% Complex variables of three complex planes of interest:
%     zp = xp + i*yp ==> Circle plane
%     z = x + i*y   ==> Intermediate plane
%     w = u + i*v   ==> Airfoil (or physical) plane
clear;clc
% Step 1: Select the parameters that define the airfoil of interest.
% (1) Select the a == angle of attack alpha
        a = 2;        % in degrees
        a = a*pi/180; % Conversion to radians
% (2) Select the parameter related to thichkness of the airfoil:
        e = .1;
% (3) Select the shift of y-axis related to camber of the airfoil:
        f = .1;
% (4) Select the trailing edge angle parameter:
        te = .05;        % 0 < te < 1 (0 ==> cusped trailing edge)
        n = 2 - te;      % Number related to trailing edge angle.
        tea = (n^2-1)/3; % This is a Karman-Trefftz extension.
% Step 2: Compute the coordinates of points on circle in zp-plane:
        R = 1 + e;
        theta = 0:pi/200:2*pi;
        yp = R * sin(theta);
        xp = R * cos(theta);
% Step 3: Transform coordinates of circle from zp-plane to z-plane:
        z = (xp - e) + i.*(yp + f);
% Step 4: Transform circle from z-plane to airfoil in w-plane
% (the w-plane is the "physical" plane of the airfoil):
        rot = exp(-i*a); % Application of angle of attack.
        w = rot .* (z + tea*1./z); % Joukowski transformation.
% Step 5: Plot of circle in z-plane on top of airfoil in w-plane
        plot(xp,yp), hold on
        plot(real(w),imag(w),'r'),axis image, hold off
```

14.9 OTHER NUMERICAL METHODS

The ODEs considered earlier in this chapter are all *initial value* problems. For *boundary value* problem solvers, see **MATLAB Help: Mathematics: Differential Equations: Boundary Value Problems for ODEs.**

MATLAB has a large number of functions for handling other numerical procedures, such as curve fitting, correlation, interpolation, minimization, filtering and convolution, and (fast) Fourier transforms. Consult **MATLAB Help: Mathematics: Polynomials and Interpolation** and **Data Analysis and Statistics.**

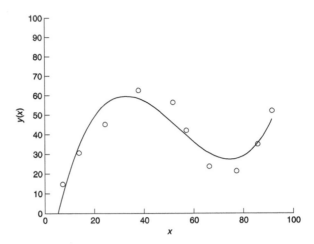

FIGURE 14.9
A cubic polynomial fit.

Here's an example of curve fitting. The following script enables you to plot data points interactively. When you have finished plotting points (signified when the *x* coordinates of your last two points differ by less than 2 in absolute value) a cubic polynomial is fitted and drawn (see Fig. 14.9).

```
% Interactive script to fit a cubic to data points

clf
hold on
axis([0 100 0 100]);

diff = 10;
xold = 68;
i = 0;
xp = zeros(1);        % data points
yp = zeros(1);

while diff > 2
  [a b] = ginput(1);
  diff = abs(a - xold);
  if diff > 2
    i = i + 1;
    xp(i) = a;
    yp(i) = b;
    xold = a;
    plot(a, b, 'ok')
  end
end
```

```
p = polyfit(xp, yp, 3 );
x = 0:0.1:xp(length(xp));
y= p(1)*x.^3 + p(2)*x.^2 + p(3)*x + p(4);
plot(x,y), title( 'cubic polynomial fit'), ...
    ylabel('y(x)'), xlabel('x')
hold off
```

Polynomial fitting may also be done interactively in a figure window, with **Tools -> Basic Fitting**.

SUMMARY

- A numerical method is an approximate computer method for solving a mathematical problem which often has no analytical solution.
- A numerical method is subject to two distinct types of error: rounding error in the computer solution, and *truncation error*, where an infinite mathematical process, like taking a limit, is approximated by a finite process.
- MATLAB has a large number of useful functions for handling numerical methods.

EXERCISES

14.1 Use Newton's method in a script to solve the following (you may have to experiment a bit with the starting values). Check all your answers with `fzero`. Check the answers involving polynomial equations with `roots`.

Hint: use `fplot` to get an idea of where the roots are, e.g.,

```
fplot('x^3-8*x^2+17*x-10', [0 3])
```

The Zoom feature also helps. In the figure window select the Zoom In button (magnifying glass) and click on the part of the graph you want to magnify.
 (a) $x^4 - x = 10$ (two real roots and two complex roots)
 (b) $e^{-x} = \sin x$ (infinitely many roots)
 (c) $x^3 - 8x^2 + 17x - 10 = 0$ (three real roots)
 (d) $\log x = \cos x$
 (e) $x^4 - 5x^3 - 12x^2 + 76x - 79 = 0$ (four real roots)

14.2 Use the Bisection method to find the square root of 2, taking 1 and 2 as initial values of x_L and x_R. Continue bisecting until the maximum error is less than 0.05 (use Inequality (14.2) of Section 14.1 to determine how many bisections are needed).

14.3 Use the Trapezoidal rule to evaluate $\int_0^4 x^2 dx$, using a step-length of $h = 1$.

14.4 A human population of 1000 at time $t = 0$ grows at a rate given by

$$dN/dt = aN,$$

where $a = 0.025$ per person per year. Use Euler's method to project the population over the next 30 years, working in steps of (a) $h = 2$ years, (b) $h = 1$ year and (c) $h = 0.5$ years. Compare your answers with the exact mathematical solution.

14.5 Write a function file `euler.m` which starts with the line

```
function [t, n] = euler(a, b, dt)
```

and which uses Euler's method to solve the bacteria growth DE (14.8). Use it in a script to compare the Euler solutions for $dt = 0.5$ and 0.05 with the exact solution. Try to get your output looking like this:

time	dt = 0.5	dt = 0.05	exact
0	1000.00	1000.00	1000.00
0.50	1400.00	1480.24	1491.82
1.00	1960.00	2191.12	2225.54
...			
5.00	28925.47	50504.95	54598.15

14.6 The basic equation for modeling radio-active decay is

$$dx/dt = -rx,$$

where x is the amount of the radio-active substance at time t, and r is the decay rate.

Some radio-active substances decay into other radio-active substances, which in turn also decay. For example, Strontium 92 ($r_1 = 0.256$ per hr) decays into Yttrium 92 ($r_2 = 0.127$ per hr), which in turn decays into Zirconium. Write down a pair of differential equations for Strontium and Yttrium to describe what is happening.

Starting at $t = 0$ with 5×10^{26} atoms of Strontium 92 and none of Yttrium, use the Runge-Kutta method (`ode23`) to solve the equations up to $t = 8$ hours in steps of $1/3$ hr. Also use Euler's method for the same problem, and compare your results.

14.7 The springbok (a species of small buck, not rugby players!) population $x(t)$ in the Kruger National Park in South Africa may be modeled by the equation

$$dx/dt = (r - bx \sin at)x,$$

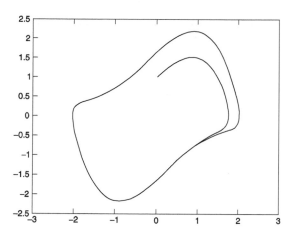

FIGURE 14.10
A trajectory of Van der Pol's equation.

where r, b, and a are constants. Write a program which reads values for r, b, and a, and initial values for x and t, and which uses Euler's method to compute the impala population at monthly intervals over a period of two years.

14.8 The luminous efficiency (ratio of the energy in the visible spectrum to the total energy) of a black body radiator may be expressed as a percentage by the formula

$$E = 64.77T^{-4} \int_{4\times10^{-5}}^{7\times10^{-5}} x^{-5}(e^{1.432/Tx} - 1)^{-1}dx,$$

where T is the absolute temperature in degrees Kelvin, x is the wavelength in cm, and the range of integration is over the visible spectrum. Write a general function `simp(fn, a, b, h)` to implement Simpson's rule as given in Equation (14.4).
Taking $T = 3500$ K, use `simp` to compute E, firstly with 10 intervals ($n = 5$), and then with 20 intervals ($n = 10$), and compare your results. (Answers: 14.512725% for $n = 5$; 14.512667% for $n = 10$)

14.9 Van der Pol's equation is a second-order non-linear differential equation which may be expressed as two first-order equations as follows:

$dx_1/dt = x_2$
$dx_2/dt = \epsilon(1 - x_1^2)x_2 - b^2x_1.$

The solution of this system has a stable limit cycle, which means that if you plot the phase trajectory of the solution (the plot of x_1 against x_2) starting at any point in the positive x_1-x_2 plane, it always moves continuously into the same closed loop. Use `ode23` to solve this system

numerically, for $x_1(0) = 0$, and $x_2(0) = 1$. Draw some phase trajectories for $b = 1$ and ϵ ranging between 0.01 and 1.0. Fig. 14.10 shows you what to expect.

Signal Processing

THE OBJECTIVE OF THIS CHAPTER IS TO INTRODUCE SIGNAL PROCESSING. YOU WILL LEARN ABOUT TWO METHODS, OF MANY, TO EXAMINE LARGE DATA SETS. THEY ARE:

- Harmonic analysis.
- Fast Fourier Transform analysis.

CONTENTS

Harmonic analysis........... 330

Fast Fourier Transform (FFT) 335

In this chapter we apply MATLAB to examine periodic data sets. Harmonic analysis is introduced in the next section to examine periodic data sets. In the subsequent section the fast Fourier transform (FFT) method is applied to examine the temporal-evolution of measured or computed data. These procedures are important methods in the field of signal processing.

A *signal* is a sequence of data that is measured by experimental methods or computed by computational methods. For example, the data $f(x)$ is measured at relatively small intervals of a particular variable x, e.g., x could be time or a space dimension. If the signal is periodic in a range of x, e.g., $-\pi$ to π, harmonic analysis is a useful and quite practical technique. The FFT method is a related method. Both methods are associated with the fitting of data with a Fourier series. In addition, both methods are widely used in engineering, e.g., in naval architecture, acoustics (or noise) and vibration engineering, electrical engineering, ocean engineering, etc.

The theory of *Fourier series* is an important topic in mathematics for physicists and engineers. A reasonably thorough discussion of this topic can be found in *Mathematics of Physics and Modern Engineering*, Second Edition, by I.S. Sokolnikoff and R.M. Redheffer, a book published by McGraw-Hill (1966). Because of the importance of this topic, it should not be surprising that there are many other books that describe the theory behind the method and the practical applications of the method. In this introductory section we give the results associated with applying the Fourier series method that are needed to examine signals.

329

Essential MATLAB for Engineers and Scientists. https://doi.org/10.1016/B978-0-08-102997-8.00022-1

Let us suppose $f(x)$ is given by the *trigonometric series*

$$f(x) = \frac{1}{2}a_o + \sum_{n=1}^{\infty}(a_n \cos nx + b_n \sin nx) \tag{15.1}$$

where $-\pi \leq x \leq \pi$ and, hence,

$$a_n = \frac{1}{\pi}\int_{-\pi}^{\pi} f(x) \cos nx \, dx \tag{15.2}$$

for $n = 0, 1, 2, \ldots$, and

$$b_n = \frac{1}{\pi}\int_{-\pi}^{\pi} f(x) \sin nx \, dx \tag{15.3}$$

for $n = 1, 2, 3, \ldots$. This assumes that the function $f(x)$ is periodic, i.e., $f(x) = f(x + 2\pi)$.

If we are given a data set $f(x)$ and it is assumed to be periodic in the interval $-\pi \leq x \leq \pi$, we can compute the coefficients a_n and b_n by applying the above formulas. With this approach we say that we expanded $f(x)$ in or say that $f(x)$ has the *Fourier series*

$$f(x) \sim \frac{1}{2}A_o + \sum_{n=1}^{N}(A_n \cos nx + B_n \sin nx) \tag{15.4}$$

where N is a positive integer greater than or equal to 1. It is the number of terms to be evaluated to construct a Fourier series (or Fourier expansion) of $f(x)$. If $N \to \infty$ and $f(x)$ is continuous and periodic, the last equation is essentially the trigonometric expansion of $f(x)$ and the result is exact. An important advantage of this series is that it can represent discontinuous functions as well. Of course, in this case or in cases where the trigonometric representation is not exact, it is a best fit representation of $f(x)$ in terms of the assumed periodic function, Eq. (15.4).

15.1 HARMONIC ANALYSIS

Let us examine the data illustrated in Fig. 15.1 by fitting a trigonometric series through the data. This sample set of "observed" data was generated by using the following formula:

$$f = .6 - \cos x - .5 \cos 2x + .4 \cos 3x + \sin x + \text{error}, \tag{15.5}$$

where the error was generated by the random number generator `randn`, a built-in function in MATLAB. Eq. (15.5) was selected for this example because it is

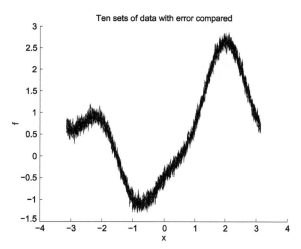

FIGURE 15.1
Sample of ten repeated periodic data sets with error.

a five-term trigonometric serious plus an error term. It was selected so that we can examine how well harmonic analysis plucks out the harmonics (i.e., the first five terms) from a signal that contains error (the error in some applications of signal processing is also called "noise").

We want to fit a finite trigonometric sum to a set of observed data. In this case the discrete set of observed data contains the values of $f(x)$ at $x = -\pi$, $-\pi + \pi/N$, $-\pi + 2\pi/N$, ..., $\pi - \pi/N$. It is assumed that $f(-\pi) = f(\pi)$. Note that $-\pi \leq x < \pi$ is the domain of x that the data are observed. The values of $f(x)$ are evaluated (or observed) at equally spaced intervals of $dx = \pi/N$. The example numerical analysis is for $N = 200$ where $N_p = 2N + 1$ is the number of equally spaced values of f from 0 to π where the data are measured. The data obtained at these points were computed by applying Eq. (15.5). Since f is prescribed on the interval $-\pi \leq x \leq \pi$ and it is assumed that $f(-\pi) = f(\pi)$, harmonic analysis is a good way to examine the data to determine its *harmonic content*, i.e., to determine the coefficients of the Fourier series that we are applying to expand the data. Note that the noise in the data introduced by the random number generator introduces an error in the periodicity assumption. Hence, in the example below ten sets of the observed data are examined by repeating the computational experiment ten times. The results illustrate that the error (or noise), in this case, is random and can be averaged away if a sufficient number of the same experiment is performed.

It turns out that if you have N_p equally spaced data points we can fit the data with $N_p - 1$ terms of a Fourier series. The results are exact in the sense that we have a system of linear $N_p - 1$ equations in $N_p - 1$ unknowns. It is reasonably

well known (see, e.g., the book by Sokolnikoff and Redheffer cited above) that the solution to this problem is

$$f = A_o + \sum_{n=1}^{N} A_n \cos nx + \sum_{n=1}^{N-1} B_n \sin nx \tag{15.6}$$

(Note that A_o is twice the value of a_o in the Fourier series given in the introduction to this chapter.) In this case, it can be shown that

$$A_o = \frac{1}{2N} \sum_{m=1}^{2N} f\left(-\pi + \frac{(m-1)\pi}{N}\right) \tag{15.7}$$

$$A_N = \frac{1}{2N} \sum_{m=1}^{2N} f\left(-\pi + \frac{(m-1)\pi}{N}\right) \cos\left((m-1)\pi\right) \tag{15.8}$$

$$A_n = \frac{1}{N} \sum_{m=1}^{2N} f\left(-\pi + \frac{(m-1)\pi}{N}\right) \cos\frac{(m-1)n\pi}{N}, \quad n = 1, 2, 3, \ldots, N-1 \tag{15.9}$$

$$B_n = \frac{1}{N} \sum_{m=1}^{2N} f\left(-\pi + \frac{(m-1)\pi}{N}\right) \sin\frac{(m-1)n\pi}{N}, \quad n = 1, 2, 3, \ldots, N-1 \tag{15.10}$$

This system of equations was applied to examine the data generated by Eq. (15.5). The MATLAB script developed and applied is as follows:

```
% Harmonic analysis
clear;clc
for ir = 1:10
N = 200;    % Nc = N+1 implies x = 0 (center of domain).
            % Np = 2N+1 are the number of points on the
            % closed interval x = [-pi, pi].
dx = pi/N; % Spacing between the points on x.
 x = -pi:dx:pi;
 f = [.6 - cos(x) - .5*cos(2*x) + .4*cos(3*x) ...
     + sin(x) + 0.1*randn(1,length(x))];
AO = (1/2/N) * sum(f(1:end-1));
A(N) = (1/2/N) * sum(f(1:end-1).*cos(N*x(1:end-1)));
  for jn = 1:N-1
        A(jn) = (1/N)*sum(f(1:end-1).*cos(jn*x(1:end-1)));
        B(jn) = (1/N)*sum(f(1:end-1).*sin(jn*x(1:end-1)));
  end
```

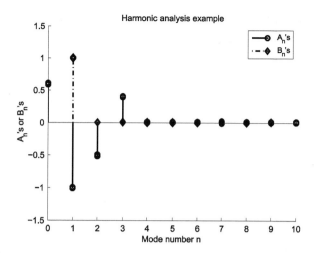

FIGURE 15.2
Harmonic analysis of the ten periodic data sets, f, compared.

```
 AA = [A0 A];
figure(1)
hold on
stem([0:length(AA(1:11))-1],AA(1:11),'ko','LineWidth',2)
stem(B(1:9),'-.kd','LineWidth',2)
legend('A_n''s', 'B_n''s')
xlabel('Mode number n')
ylabel('A_n''s or B_n''s')
title('Harmonic analysis example')
hold off
figure(2)
hold on
plot(x,f,'k')
title('Ten sets of data with error compared')
xlabel('x'),ylabel('f')
hold off
end
```

Note that the algorithm was executed ten times to examine the influence of the random error on a converged solution. The random error is plus-or-minus 10% of unity. The ten samples of the periodic data examined in the script is illustrated in Fig. 15.1. The ten associated harmonic-analysis results are given in Fig. 15.2. As we might have expected, the results, on average, are pretty close to the exact results described next.

Since the function $f(x)$ without error is

$$f = .6 - \cos x - .5 \cos 2x + .4 \cos 3x + \sin x$$

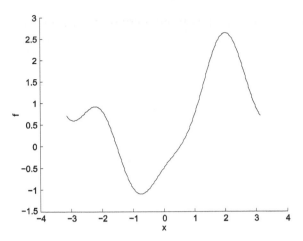

FIGURE 15.3
Sample of the periodic data set, $f(x)$, without error.

FIGURE 15.4
Harmonic analysis of the periodic data, $f(x)$, without error.

it is a five term trigonometric series. The harmonic analysis of this function is exact. This function and the exact results of the harmonic analysis of it are illustrated in Fig. 15.3 and Fig. 15.4, respectively. The figures were generated with the same MATLAB script given above with the following changes. The coefficient of the error term was set to zero as follows: We changed `0.1*randn(1,length(x))];` to `0*randn(1,length(x))];`. Since the ten repeated trials are identical, if you change the repeated-trials `for` loop, `for ir = 1:10`, to `for ir = 1:1`, the results you obtain will be identical to the results in the figures for the zero error case.

Comparing the error-less data with the noisy data generated by adding a random error, illustrates the need for making repeated measurements in the laboratory to get a sensible result for the amplitudes of the harmonics that represent the shape of the phenomenon sought, viz., $f(x)$ in the interval $-\pi \leq x \leq \pi$. Note that if a trigonometric series is used to define the input data set, the output of harmonic analysis yields an exact reproduction of the input series.

15.2 FAST FOURIER TRANSFORM (FFT)

James W. Cooley and John W. Tukey published *An Algorithm for the Machine Calculation of Complex Fourier Series* in 1965; it appeared in *Mathematics of Computations* **19**, 297–301. This paper was cited in the book edited by Leo L. Beranek and István L. Vér titled *Noise and Vibration Control Engineering* published by John Wiley & Sons (1992). This book provides a good review of signal processing as well as its practical applications in noise and vibration engineering. The discussion to follow follows closely the discussion in this book.

The Fourier transform of a time history signal, $y(t)$, that is measured over the time interval $0 \leq t \leq T$ is defined for all frequencies, both positive and negative, by

$$Y(\omega, T) = \int_0^T y(t) e^{-i2\pi \omega t}\, dt \tag{15.11}$$

where Y is a function of frequency, ω. In terms of a digital time series of N data values where $y(t) = y(n\,\Delta t)$, $n = 0, 1, 2, \ldots, N - 1$. The Fourier transform may be written as a complex Fourier series

$$Y(\omega, T) = Y(k\Delta\omega, N) = \Delta t \sum_{n=0}^{N-1} y(n\,\Delta t)\, exp(-i2\pi \omega n \Delta t) \tag{15.12}$$

where the spectral components, i.e., the $Y(k\Delta\omega, N)$'s, are generally complex valued and are defined only at N discrete frequencies

$$\omega_k = k\Delta\omega = \frac{k}{N\,\Delta t}, \quad k = 0, 1, 2, \ldots, N - 1$$

Hence, this is essentially the complex version of the Fourier series expansion introduced in the introduction of this chapter. To clarify this point further, let

us divide Eq. (15.12) by $T = N\Delta t$. This yields the following result:

$$C_k = \frac{Y(k\Delta\omega, N)}{N\Delta t} = \frac{1}{N}\sum_{n=0}^{N-1} x(n\,\Delta t)\,exp(-i2\pi\omega n\Delta t)$$

$$= \frac{1}{N}\sum_{n=0}^{N-1} x(n\,\Delta t)\,exp(-i2\pi k n/N) \tag{15.13}$$

Note the analogy between this formula and the formulas for the Fourier coefficients given in the previous section on harmonic analysis. In this case, however, the coefficients are complex numbers. Since MATLAB deals with complex numbers and does complex arithmetic, applying the FFT methodology to examine signals is reasonably straightforward. This will be illustrated below by an example.

The complex Fourier coefficients defined by Eq. (15.13) are for an assumed periodic function and, hence, the time history $y(t)$ is assumed to repeat over periods equal to the sampling period, T. The Fourier components are unique only out to $k = N/2$, i.e., out to the frequency $\omega_k = 1/(2\,\Delta t)$, commonly known as the Nyquist frequency, ω_N, of the digital signal. At this frequency there are only two sample values per cycle and, hence, the error known as *aliasing* is initiated; for details see, e.g., Beranek and Vér (1992) cited above. The first $N/2 + 1$ Fourier coefficients, from $k = 0$ to $k = N/2$, define the spectral components (i.e., magnitudes of the Fourier coefficients of significant magnitude) at positive frequencies, while the last $N/2 - 1$ Fourier coefficients, from $k = N/2 + 1$ to $k = N - 1$, define the spectral components at negative frequencies.

The various FFT algorithms developed since the publication of *Cooley-Tukey* algorithm are well documented in the technical literature. The details of developing and FFT algorithm is also beyond the scope of this presentation. However, since the procedure applied in the example examined in this section can be applied to other problems that may be of interest to the reader, a few of the hints that need to be considered in applying the Cooley-Tukey algorithm pointed out by Beranek and Vér are worth noting since this algorithm is the basis for the FFT algorithm applied in the example discussed in this section. They are:

- It is useful to restrict the number of equally spaced data points for each FFT to a power of 2, i.e., $N = 2^p$, where $p = 8$ to 12 are commonly used.
- The frequency resolution of the Fourier components will be $1/(N\,\Delta t)$.
- The Nyquist frequency where aliasing is initiated occurs at $k = N/2$ and it is $\omega_N = 1/(2\,\Delta t)$. The sampling frequency should be at least twice the highest frequency expected to be found in the signal.

- The first $N/2 + 1$ Fourier components up to the Nyquist frequency are related to the last $N/2 - 1$ components above the Nyquist frequency by $Y(k) = Y^*(N - k)$, $k = 0, 1, 2, \ldots, N - 1$, where the asterisk denotes complex conjugate.
- The Fourier components that define only the positive frequencies is called a one-sided spectrum. This spectrum is given by $Y(0)$, $Y(N/2)$, and $2Y(k)$, $k = 1, 2, \ldots, N/2 - 1$.

The one-sided spectral components for a periodic signal with zero mean are given by

$$P(\omega) = \frac{2Y(\omega, T)}{T}, \quad \omega > 0, \quad \text{or} \quad k = 1, 2, \ldots, \frac{N}{2} - 1 \tag{15.14}$$

where $Y(\omega, T)$ is defined in Eq. (15.12). The magnitudes of the Fourier components, $|P(\omega)|$, are usually plotted as a `stem` plot (see example below) and, hence, it is often called a line spectrum. Because each of the Fourier components are complex, they define both a phase and an amplitude for each of the components. However, the phase information is generally utilized only in applications where there may be a need to reconstruct the time history of the signal or to determine the peak values of the signal.

Let us consider the following example to illustrate how MATLAB can be used to apply an FFT to a signal. Consider the following signal:

$$y(t) = 0.7\sin(2\pi 50t) + \sin(2\pi 120t) + \text{noise} \tag{15.15}$$

where the sample is taken over the interval of time $0 \le t \le T$. The sampling frequency is `Fs = 3000`. Thus $T = 1/Fs$. The power spectrum is also usually plotted to isolate the dominant frequencies in a signal. It is `Pyy` $= (Y(\omega, T)Y^*(\omega, T))/L$, where $L = 2^p$ and $p = 11$ was selected in this example. With this brief description of a signal processing problem, the following script was applied that implements the FFT methodology to examine this signal (it is based on an example in the `help` documentation available within MATLAB).

```
% ---------Example of the application of FFT------------
% Example of the application of FFT
Fs = 3000;                    % Sampling frequency
T = 1/Fs;                     % Sample time
pwr2 = 11;
L = power(2,pwr2);            % Length of signal
t = (0:L-1)*T;               % Time vector for FFT
% Sum of a 50 Hz sinusoid and a 120 Hz sinusoid
x =   0.7*sin(2*pi*50*t) + sin(2*pi*120*t);
y = x + 2*randn(size(t));    % Sinusoids plus noise at t
```

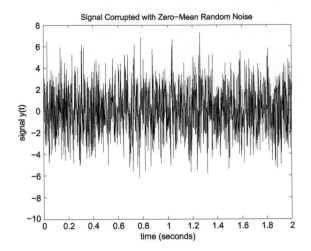

FIGURE 15.5

Example of signal to be analyzed by spectral analysis.

```
plot(Fs*t/1000,y,'k'),axis([0 2 -10 8])
title('Signal Corrupted with Zero-Mean Random Noise')
xlabel('time (seconds)'),ylabel('signal y(t)')
figure(2)
NFFT = 2^nextpow2(L); % Next power of 2 from length of y
 Y = fft(y,NFFT)/L;
 Pyy = Y.*conj(Y)/L;
 f = Fs/2*linspace(0,1,NFFT/2+1);
%
% ---------Plot single-sided amplitude spectrum.--------
FF = 2*abs(Y(1:NFFT/2+1));
stem(f(1,1:110),FF(1,1:110),'k','Linewidth',2)
title('Single-Sided Amplitude Spectrum of y(t)')
xlabel('Frequency (Hz)')
ylabel('|Y(\omega)|')
figure(3)
stem(f(1:110),Pyy(1:110),'k','Linewidth',2)
title('Power spectral density')
xlabel('Frequency (Hz)')
ylabel('Pyy')
```

The signal to be analyzed by spectral analysis (as it is called) is illustrated in Fig. 15.5. The line spectrum and the power spectrum predicted for this signal are given in Fig. 15.6 and Fig. 15.7, respectively. There are two things to note. One is that the dominant frequencies in the signal are identified clearly in the spectra. The second is that the power spectrum isolates the dominant modes in the signal more clearly.

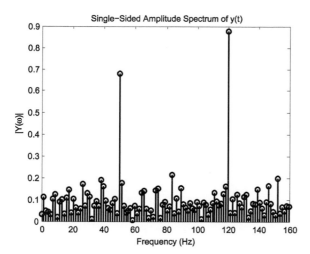

FIGURE 15.6
Line spectrum of the signal in Fig. 15.5.

FIGURE 15.7
Power spectrum of the signal in Fig. 15.5.

SIMULINK® Toolbox

THE OBJECTIVE OF THIS CHAPTER IS TO INTRODUCE SIMU-LINK BY APPLYING IT TO SOLVE DYNAMIC SYSTEMS INVES-TIGATED BY SCIENCE, TECHNOLOGY, ENGINEERING AND MATHEMATICS STUDENTS.

The first topic covered in this chapter is the process of opening Simulink and initiating the construction of a signal processing model. Subsequently, we examine two examples of dynamic systems typically investigated by students in their first-year courses in physics. The next two dynamic systems examined are classic nonlinear equations typically investigated in courses on nonlinear dynamical systems. The examples examined in this chapter are as follows:

- An example of signal processing illustrating the initiation of the construction of a Simulink model.
- A spring, mass damper model of a mechanical system.
- A bouncing ball model.
- The van der Pol oscillator.
- The Duffing oscillator.

CONTENTS

Mass-spring-damper dynamic system 347

Bouncing ball dynamic system 349

The van der Pol oscillator 352

The Duffing oscillator 353

Exercises 355

In this chapter an introduction to applying Simulink software via relatively simple yet interesting examples from classical mechanics is presented. In a typical science, technology, engineering and mathematics program at universities in the US the students are introduced to dynamical systems modeling in their first-year course on physics. In the first course they learn about systems that can be modeled by springs, masses and damping devices. In the second course the analogous electrical system that can be modeled by resistors, inductors and capacitors is investigated. How to simulate the dynamics of these systems with Simulink is examined in this chapter.

What is Simulink? As described in the help available with this toolbox: It is used to model, simulate and analyze dynamic systems. It enables you to pose questions about a system, model the system and observe what happens. "With Simulink, you can easily build models from scratch, or modify existing models to meet your needs. Simulink supports linear and nonlinear systems, modeled in continuous time, sampled time, or a hybrid of the two. Systems can also

341

Essential MATLAB for Engineers and Scientists. https://doi.org/10.1016/B978-0-08-102997-8.00023-3

FIGURE 16.1
Simulink window start page.

be multirate—having different parts that are sampled or updated at different rates." It is a graphical programming environment.

To conclude this introduction we will examine a simple example of a signal processing model. Recall that the help files available can be opened by clicking the question mark, '?', just below the word "Help" in the toolbars at the top of the MATLAB desktop. Click on Simulink to open the e-manual on this toolbox. The guidelines provided in the help for the development of more sophisticated dynamic system models provides one of the key suggestions for learning to use Simulink. It is to develop a new model by starting, if possible, with an existing model, modifying it to develop a new model to solve a new problem.

To start the exercises we need to open Simulink. This can be done by executing the following command in the command window.

```
>> simulink
```

This command opens the Simulink Start Window illustrated in Fig. 16.1. Click on the Blank Model pain near the middle of the start window. A model building window illustrated in Fig. 16.2 opens. Click the icon in the toolbar to open the library browser. This browser is illustrated in Fig. 16.3. In the left panel of the browser an index of the Libraries is provided. Pointing and clicking on any of the items opens up the various utilities available to design (or develop) a model within Simulink to simulate a dynamic system. By single-clicking on the "white-page" icon below the word "File" just below the title of the browser window opens a new model; this is illustrated in Fig. 16.3. Pull down the File menu in the untitled model window, point and click "save" and name the

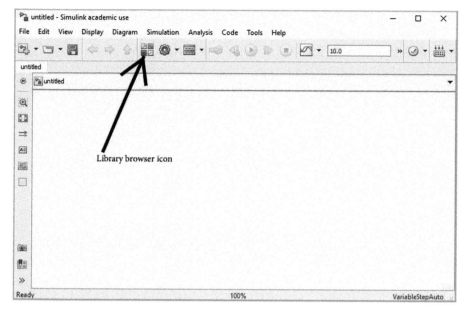

FIGURE 16.2
The Simulink untitled coding window.

FIGURE 16.3
The Simulink library browser with the Simulink untitled coding window.

model that you plan to build, e.g., name it Example1. This operation creates a file named Example1.mdl. Note the file extension .mdl. This is the file extension that identifies the file as a Simulink model. The new model window with this

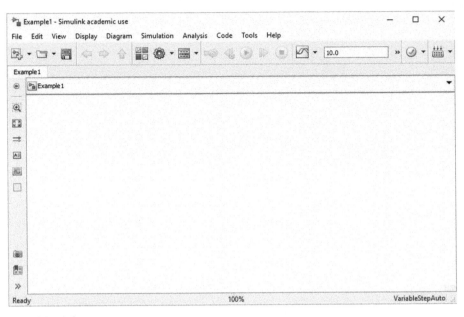

FIGURE 16.4
The untitled working window renamed "Example1".

name is illustrated in Fig. 16.4. This will help save your work as you build and debug the model to solve a particular dynamic system problem.

While in the model window hold "Ctrl" and "e" simultaneously to open the solver configuration window illustrated in Fig. 16.5. Note that the default "Relative tolerance" is 1e-3, the "Solver" is ode45 among other parameters. Except for the first example below the "Relative tolerance" was changed to 1e-6. In addition, in some cases the solver applied was changed. The solver applied is also noted at the bottom right of the model window.

Close the model by clicking the x in the upper right-hand corner of the model window. To open this file again double-click on it in the Current Folder window on the MATLAB desktop. We will build a simple model in this file window in the beginning of the next section of this chapter.

It is expected, of course, that the reader reproduce the next example on data analysis and subsequent examples on dynamic-system modeling to study the application of Simulink to solve technical computing problems. The examples are relatively simple; they are similar to the dynamic system problems examined in Chapters 14 and 17 with Matlab codes.

To illustrate the mechanics of constructing a model let us consider a signal-processing example. In this example we are going to examine the output from a sine-wave generator. We will integrate the area under the sine wave and compare this with the original signal on a scope, a device that plots the sine wave

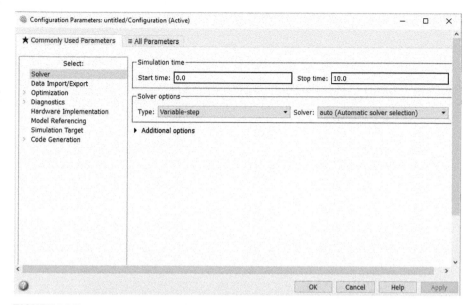

FIGURE 16.5
Simulink window illustrating the solver configuration; this is opened by clicking the icon to the right of the Library Browser icon.

and the integral of the sine wave with respect to time from time equal to zero, i.e., the time the simulation is started, to the specified time (in this case the default is 10). This model was built by applying the following steps.

1. Point-and-click on the "Sources" icon in the Simulink Library Browser. This opens a palette from which we want to drag the "Sine Wave" and drop it onto the `Example1` window illustrated in Fig. 16.4 (we are trying to construct the model illustrated in Fig. 16.6). This sine wave generator is what we need to provide the input to the model we are building in this section.

2. Next, open up the "Commonly Used Blocks" by scrolling up the contents and pointing-and-clicking on this topic to open up another palette. Find on this palette a "Scope", an "Integrator", and a "Mux". Next, drag and drop each of these devices onto the `Example1` window.

3. In the `Example1` window drag the cursor to the right-side output port of the "Sine Wave" generator. When you see the cross, hold the left-mouse button down and drag the cross to the top-left input port of the "Mux". This is one way to connect a "wire" from one device to the next.

4. Following the same procedure connect the right-side port of the "Mux" to the left-side input port of the "Scope".

5. Starting at the left-side input port of the "Integrator" hold the left-mouse button down and move to any location on the wire already connecting the "Sine Wave" generator to the "Mux".

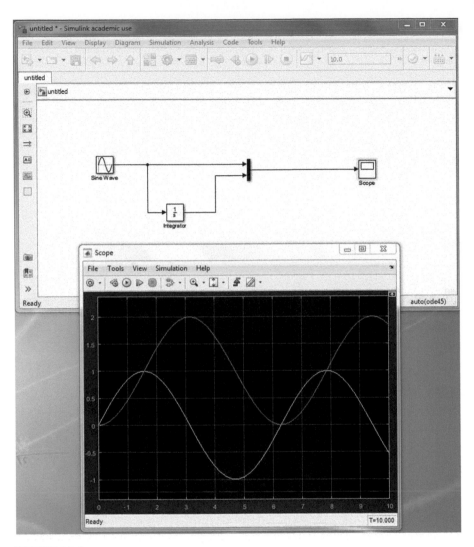

FIGURE 16.6
The Example1 model and the scope illustrating the results of this simulation.

6. Finally, connect the right-side output port of the "Integrator" to the lower-left input port of the "Mux".
7. Pull down the file menu and click "save" to save your work. In this example we already saved the name of this file, viz., `Example1.mdl`.
8. To continue, double click on the "Scope" to open the Scope window.
9. Finally, click on the model window to make sure it is in the forefront. To start the simulation hold the keys "Ctrl" and "T" down simultaneously.

You should see two curves appear on the scope. It will be the sine wave (yellow) and the integral of the sine wave (magenta). The "Mux" allows for two inputs into the scope. The model and the results of its execution are illustrated in Fig. 16.6.

Of course, if you haven't had any experience with experimental equipment like oscilloscopes, sine-wave generators, multiplexers and the like, much of this may seem a bit mysterious. In addition, the exercise above is only a look-see exercise. It hardly demonstrates the many more powerful features available within Simulink. However, more experience in your technical training and education will allow you to explore more productively this powerful tool and clear up many of these matters. In addition, learning more about Simulink by exercising many of the examples described in the help documentation will help to understand the other toolboxes, e.g., the Control Systems Toolbox. Furthermore, in courses on Dynamical Systems in mechanical and in electrical engineering, many of the available textbooks ask the student to apply Simulink to investigate the systems examined. Thus, there are numerous sources with sample applications, including searching the computer web, that should prove beneficial in learning how to apply this tool to solve more substantial problems in engineering and science. Additional details on the various utilities applied in the above example can be found in the help. We next examine one of the simplest yet quite useful dynamic system.

16.1 MASS-SPRING-DAMPER DYNAMIC SYSTEM

The simplest ordinary differential equation that can be used to investigate the oscillation of a dynamic system is the following differential equation:

$$\frac{d^2x}{dt^2} + b\frac{dx}{dt} + \omega_o^2 x = A\sin(\omega t). \tag{16.1}$$

The equation can be used to describe the motion of a mechanical system modeled by a mass, spring and damper forced by a sinusoidal forcing function, where $b = R/m$ is the resistance coefficient divided by the mass, m, of a lumped object, $\omega_o^2 = k/m$ is the angular frequency of oscillation of the undamped ($R = 0$) motion with a spring-restoration force coefficient k. This is, of course, Newton's second law of motion applied to the forced motion of a dynamic system modeled as a lumped mass acted on by an imposed harmonic force with amplitude $A = F/m$, where F is the amplitude of force applied to the system. This is one of the first problems typically investigated in a course on dynamics when examining the oscillatory motion of a particle in one dimension. Let us rewrite this equation as follows:

$$\frac{d^2x}{dt^2} = -b\frac{dx}{dt} - \omega_o^2 x + A\sin(\omega t). \tag{16.2}$$

FIGURE 16.7
The MassSpringDamperEx1.mdl model, the X-Y plotter and the scope illustrating the results of this simulation are illustrated in this figure.

To solve a differential equation requires integration. Since this is a second or-der equation we need to integrate twice to obtain a solution. A Simulink model to solve this equation is illustrated in Fig. 16.7. Note that the input to "Inte-grator1" is the sum of the right-hand side of the above equation, i.e., it is the value of the acceleration, d^2/dt^2. The output is the speed, dx/dt, which is the input to "Integrator2". Finally, the output of "Integrator2" is the solution at one time step. The details of the simulation illustrated in the figure is described next.

The case illustrated is for $\omega_o^2 = k/m = 1$ and $b = R/m = 0.5$. The initial position x and the initial velocity dx/dt are equal to zero. In addition, the force on the system at $t = 0$ is zero. The results indicate that for the forcing frequency of $\omega = 1$ radian per unit time and amplitude $A = F/m = 1$, the ultimate ampli-tude of the motion of the mass is about 2. There is an approximate phase lag of about $\pi/2$ between the response and the forcing. Since the forcing frequency is very nearly at the resonance frequency the results illustrated in the figure are

consistent with what is reported in the literature, e.g., in Becker (1953).[1] As an exercise vary the forcing frequency by double clicking on the "Sine Wave" generator and change the number just below "Frequency (rad/sec):". Try, e.g., frequencies in the range $0.5 \leq \omega \leq 1.5$ and compare your results with the example just described. Find a text on mechanics (or dynamics) to compare your results from Simulink with the theoretical results on this problem reported in the literature.

16.2 BOUNCING BALL DYNAMIC SYSTEM

This section illustrates the application of initial conditions other than zero (as in the previous section). The model is a modification of the bouncing ball model by MathWorks (Copyright 1990–2010 The MathWorks, Inc.) described in the help; it can be found by typing "Simulation of a bouncing ball" into the search space just below the far left of the tool bar in the "? help" window. The help window can be opened by clicking the circled question mark in the upper tool bar on the Matlab desktop. Fig. 16.8 illustrates the application of initial conditions to the integrals.

This bouncing-ball model is an example of a hybrid dynamic system. A hybrid dynamic system is a system that involves both continuous dynamics, as well as, discrete transitions where the system dynamics can change and the state values can jump. The continuous dynamics of a bouncing ball is simply given by the following equation

$$\frac{d^2x}{dt^2} = -g, \tag{16.3}$$

where g is the acceleration due to gravity, x is the vertical distance of the ball after it is released from a specified height, x_o at a specified speed, $v_o = dx/dt|_o$. The ground is assumed to be at $x = 0$. Therefore, the system has two continuous states: position, x, and velocity, $v = dx/dt$.

The hybrid system aspect of the model originates from the modeling of a collision of the ball with the ground. If one assumes a partially elastic collision with the ground, then the velocity before the collision, , and velocity after the collision, v^-, can be related by the coefficient of restitution of the ball, κ, as follows:

$$v^+ = -\kappa v^-, \quad x = 0.$$

The bouncing ball therefore displays a jump in a continuous state (velocity) at the transition condition, $x = 0$. This discussion is a paraphrase of the informa-

[1] Becker, R.A. (1953): *Introduction to Theoretical Mechanics*, McGraw-Hill Book Company, NY.

FIGURE 16.8
The bouncing ball model: The velocity and position of the ball are illustrated on the scopes.

tion on this problem given in the electronically available help files that come with MATLAB.

Let us examine the application of the integrators in this example. The first integrator is labeled "Velocity". Double clicking on top of this icon with the left mouse opens a window that gives the integrator parameters that are set for this form of the integrator. The external reset is set to "rising". The initial condition source is set to "external". The "Show state port" and "Enable zero-crossing detection" are the only items checked. This is illustrated in Fig. 16.9. The integrator parameters for the integrator labeled "Position" is shown in Fig. 16.10. The top left input port of the two integrators is the input. The bottom two ports

FIGURE 16.9
The bouncing ball model: The velocity and position of the ball are illustrated on the scopes.

FIGURE 16.10
The bouncing ball model: The velocity and position of the ball are illustrated on the scopes.

allow the initial condition to change at particular intervals of time; in this example, the time to change is associated with the time at which the ball hits the ground. More details on the integrator options at your disposal is described in the help. Search for integrator within the help. As pointed out in the help, "the integrator block outputs the integral of its input at the current time step". Reading the help files and applying the various options will help you learn more about the capabilities of the Simulink utilities. Certainly there are many ways to construct models to solve a particular problem; it depends on the number of utilities that you are familiar. Even with just the few utilities applied in this and in the previous examples, a number of more complex problems can be solved. This is illustrated in the next two examples on nonlinear dynamical systems.

16.3 THE VAN DER POL OSCILLATOR

In this section we illustrate a model to solve the differential equation known as the *van der Pol* oscillator. In investigations of nonlinear differential equations and in the study of nonlinear dynamics this equation is one of the first models typically investigated. It has been used, e.g., in the analysis of a vacuum-tube circuit among other practical problems in engineering. The van der Pol equation is

$$\frac{d^2x}{dt^2} = b\left(1 - x^2\right)\frac{dx}{dt} - x. \tag{16.4}$$

The interesting thing about finding equations like this one in the engineering and scientific literature is that we can investigate its behavior by applying the technical computing capabilities in MATLAB/Simulink. In this chapter we examine the application of Simulink to solve this equation.[2]

A Simulink model of this equation is illustrated in Fig. 16.11. In this example we selected $b = 5$. Also, we had to change the initial condition for the device named "Integrator". The default initial condition of an "Integrator" is zero. To change it you need to double click on the icon and change the initial condition. In this example the initial condition in the second integrator with name "Integrator" was changed to 1. If it is zero, the solution for all time is zero. Applying an initial condition like this one moves the solution from the origin and closer to the limit cycle solution illustrated in the figure. The limit cycle is a single closed curve in the "X-Y Graph" that the system approaches for large times away from the initial condition. At time far from the initial condition the system, in this case, reaches a periodic state.

[2]More can be found by searching in the help by clicking the question mark and searching for "van der Pol". A Simulink demo is one of the items that comes up when the search term used is as quoted.

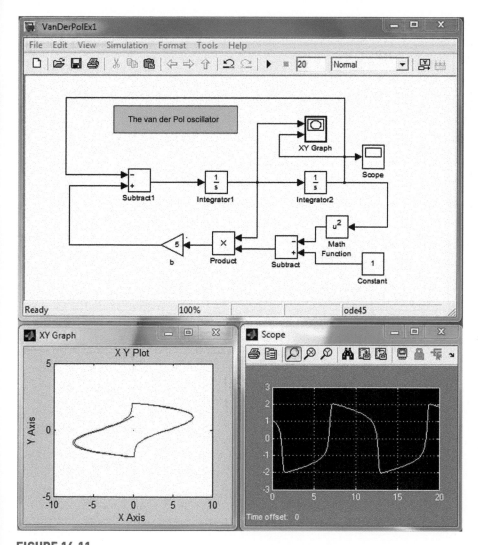

FIGURE 16.11

The van der Pol oscillator: The "Scope" illustrates the solution or output of the oscillator. The "X-Y Graph" is a plot of $X = dx/dt$ versus $Y = x$.

16.4 THE DUFFING OSCILLATOR

Fig. 16.12 illustrates the simulation of a forced nonlinear oscillator modeled by Duffing's equation, viz.,

$$\frac{d^2x}{dt^2} = Ax - Bx^3 - C\frac{dx}{dt} + F \sin \omega t. \tag{16.5}$$

FIGURE 16.12
Forced nonlinear oscillator: The "Scope" compares input forcing (yellow) and output displacement (magenta). The "X-Y Graph" is a plot of $X = dx/dt$ versus $Y = x$.

This is one of the now classical equations investigated in courses on nonlinear dynamics. In the example illustrated in the figure, $A = 1$, $B = 1$, $C = 0.22$ and $F = 0.3$. The forcing frequency is $\omega = 1$ rad/s. The latter can be verified by double clicking on the "Sine Wave" icon; this operation opens a panel with information about the output of the sine wave generator. For this set of constants the phase portrait, the X-Y graph in the figure, evolves into a limit cycle or periodic orbit called a "stable period 3 orbit" as described in detail in the book by Guckenheimer and Holmes.[3] This example, like the others presented in this chapter, are intended to illustrate that you can study the properties of

[3] Guckenheimer, John, & Philip Holmes (1983): *Nonlinear Oscillations, Dynamical Systems, and Bifurcations of Vector Fields*, Springer-Verlag, NY.

the solutions to differential equations numerically relatively easily by applying the tools in Simulink (and in MATLAB).

There is a wealth of literature on the theory and the numerics of nonlinear equations and, in particular, the Duffing and the van der Pol equations. The theoretical results reported in Guckenheimer and Holmes helped the author select the constants applied in this example. Suggestions for changes in the constants are to be examined in the exercises at the end of this chapter.

EXERCISES

16.1 Reproduce the signal processing model, i.e., first example in this chapter on the display of the sine and its integral. Execute the simulation to make sure it works (i.e., compares favorably with the results in the example). Next, do the following:

(a) Double click on the integrator and change the initial condition from 0 to 1. Apply the change by clicking the button in the lower right corner of the integrator function block. Execute the simulation. What is the maximum value of the integral? What is the minimum value of the integral?

(b) Double click on the sine wave icon and change the frequency from 1 to 2 and the amplitude from 1 to 2. Apply the changes by clicking the button in the lower right corner of the sine wave source block. Execute the simulation. Note that the frequency is doubled and the amplitude is doubled.

(c) Change the phase of the sine wave from 0 to pi/2. Make sure that the amplitude and frequency are set equal to 1. Apply the changes and execute the simulation. Note that the input function is a cosine instead of a sine function.

16.2 Reproduce the bouncing-ball model and execute it to determine that your reproduction is consistent with the example in this chapter. With a working code examine changes in the coefficient of restitution. Consider several values in the range $0.5 \leq \kappa \leq 0.9$. Also examine changes in the initial conditions for $\kappa = 0.8$. To change the value of κ you need to double click on the icon for the coefficient of restitution, change its value and apply it. The same is done to change the initial velocity or the initial position. Once you examine a few different cases explain what you found.

16.3 Reproduce the model for a forced spring-mass-damper mechanical system. Remove the "X-Y Graph", its connecting wires, the "Mux" and its connecting wires. Then reattach the "Scope" to the wire attached to the output of "Integrator2". The model is the spring-mass-damper model with a step change in the force applied at $t = 0$ as can be verified by double clicking on the "Step" icon; change the "Step time" from the default value of 1 to zero if the time the step is applied is not equal to zero.

Set the resistance coefficient R/m to zero. Leave $k/m = 1$, i.e., the spring constant coefficient. Execute the simulation. What is the frequency of oscillation? Examine $R/m = .25$, 1, 2 and 4 to determine the effect of increased damping on the solution. If you look this problem up in a physics book on mechanics it is usually under the topic "damped harmonic oscillator". If $R/m = 0$, it is the harmonic oscillator solution known as simple-harmonic motion; this should have been the solution that you found as part of the examination of the dynamic system in this problem. Without damping (i.e., $R = 0$) the solution is a sinusoidal motion with constant amplitude. The frequency of oscillation is the natural frequency, $\omega_o = \sqrt{k/m} = 1$ in this case. Thus, it is not unexpected that a cycle of oscillation occurs over a time interval of 2π; review your results to make sure that this is the case. If $\omega^2 = (R/m)^2/2$, then the motion is critically damped. This means that the dynamic system reaches its infinite time, constant valued solution without overshoot. If ω_o is less than this value, the system is said to be underdamped. If ω_o is greater than this value, the system is said to be overdamped. Setting $R/m = 2$ leads to a critically damped behavior. Describe, from your solutions for the range of R/m considered in this problem, the meaning of underdamped, critically damped and overdamped response (as reflected in the solution x) of the system for the mechanical system examined in this problem.

16.4 In the previous problem try the damping coefficient $R/m = -0.25$. Describe the results. Notice how the amplitude grows with time. Thus, with a negative resistance coefficient we have growth in the amplitude of the solution. If you set $k/m = 0$, you can investigate exponential growth. Create a problem that you provide a solution that deals with the problem of exponential growth.

16.5 Reproduce the van der Pol oscillator model. Execute the simulation to compare your results with the results in the example (this is, of course, a necessary step to check your computer code). Next, set the damping coefficient $b = 0$. Execute the simulation. The result should be simple-harmonic motion of amplitude unity because the initial condition in the second integrator was set equal to unity and period equal to 2π. Next, examine the case for $b = 1$. Finally, examine $b = 5$. Note the qualitative changes in the response function, i.e., in the plotted values of x.

16.6 Reproduce the Duffing oscillator model. Execute the simulation to compare your results with the results in the example. Set the parameters $A = B = 0$. The result is forced harmonic motion at a single frequency. Try examining the linear case, which means $B = 0$, and $-1 \leq A \leq -0.1$, which means a spring. In all cases the motion ends up being a forced harmonic motion of a single sinusoidal frequency. This is because the cubic-nonlinear term was suppressed by setting $B = 0$. (There are many other combinations that you can explore. However, be aware of the fact

that some combinations of parameters lead to *chaotic* solutions and, hence, may never settle down to a periodic state. There also may be parameter selections that lead to unstable solutions.)

Symbolics Toolbox

CONTENTS

Algebra 360
Polynomials 361
Vectors 363
Matrices 364

Calculus 368

Laplace and Z
transforms 370

Generalized
functions* 371

Differential
equations 373

Implementation
of funtool,
MuPAD and help 374
The funtool 374
*The MuPAD
notebook* and
Symbolic help* 375

Exercises 377

THE OBJECTIVE OF THIS CHAPTER IS TO INTRODUCE THE SYMBOLIC TOOLS AVAILABLE WITHIN MATLAB.

It is a useful set of tools for the application of mathematics in science and engineering. Examples of the application of the Symbolic capabilities have already been given elsewhere in this book to check analytic solutions found in the literature. In addition, predictions based on the analytic solutions were compared with predictions based on computational methods. The symbolic mathematical tools covered in this chapter are:

- Algebra: Polynomials, vectors and matrices.
- Calculus: Differentiation and integration.
- Transforms: Laplace and Z transforms.
- Generalized functions: Heaviside and Dirac.
- Systems of ordinary differential equations.
- The funtool, MuPAD and help.

With the Symbolics toolbox MATLAB can be considered the "mathematical handbook" for the students of the 21st century. In the *old days* science and engineering students were required to purchase a "standard" mathematics handbook. Today, there are very few students, if any, who own and use a mathematics handbook. The computer and tools like MATLAB have changed this situation in a positive way. However, as for the second author of this book, it has been an interesting exercise to demonstrate that MATLAB is this century's handbook. In this chapter an introduction to the application of Symbolics follows a procedure based on a review of handbook information and providing examples for the kinds of information we were expected to look up in a handbook and, hence, what we, sort of, expect students to find by applying computer tools like MATLAB.

The main problem, if it is a problem at all, in the application of the symbolic toolbox, or any other computer tool that allows the user to do symbolic analysis, is that to use the tool effectively you need to know something about the mathematical questions being raised. In other words, if you want to differ-

359

entiate the function, $f(x)$, with respect to the variable x, you need to know the meaning of a function and the meaning of differentiation or finding the derivative of f with respect to x, viz., finding df/dx.

Another example is finding the solutions of a quadratic equation. You need to know what kind of equation this is and the meaning of solving an algebraic equation of this type. Hence, if you know what to ask, MATLAB Symbolics can be applied to help you find an answer to a symbolic mathematical question.

The application of symbolics to examine mathematical questions that can be answered with the help of mathematical handbooks is addressed in this chapter to illustrate the power of the symbolic tools available in MATLAB. Certainly the tools are more powerful than the relatively simple examples presented in this chapter. Once you are adept at using this toolbox, then you can extend the application of this toolbox to examine more difficult questions that may even be beyond the capabilities of more traditional applications of applied mathematics to solve problems in science and in engineering.

The symbolic mathematical topics covered in a typical handbook are, e.g., algebra, trigonometry, calculus, integration, differentiation and differential equations. The information provided in a handbook is to help the student, engineer or scientist make progress in solving mathematical problems they confront in their attempts to solve technical problems. A solid education in science, technology, engineering and mathematics (STEM) provides the necessary background to utilize handbooks and/or tools like MATLAB in a productive way. Of course, MATLAB like handbooks need to be used regularly to gain proficiency in applying such tools as part of finding solutions. We will touch some of the topics covered in handbooks to help guide the readers in their attempts to apply the Symbolic toolbox within MATLAB. We will start with algebra including linear algebra, vector algebra and matrix algebra. Then we will examine differentiation and integration. This will be followed by investigating integral transforms and, in particular, the Laplace transform. Then we will conclude by investigating the symbolic solution of differential equations.

17.1 ALGEBRA

In Section 1.1.6 we introduced how to solve a system of linear equations. One example applied the `solve` utility available in MATLAB Symbolics. Let us extend that example by solving the simultaneous system of quadratic equations

$$x^2 + 3y = 0, \quad y^2 + 2x = 0.$$

Applying the same procedure illustrated in Section 1.1.6, we get the following results by typing and executing the following command in the Command Window. The results are also illustrated below.

```
>> syms x y
>> [x y] = solve(x^2 + 3*y, y^2 + 2*x)

x =

                                                0
                                 -(-12)^(2/3)/2
  -((3^(1/2)*(-12)^(1/3)*i)/2 + (-12)^(1/3)/2)^2/2
  -((3^(1/2)*(-12)^(1/3)*i)/2 - (-12)^(1/3)/2)^2/2

y =

                                          0
                               (-12)^(1/3)
  - (3^(1/2)*(-12)^(1/3)*i)/2 - (-12)^(1/3)/2
    (3^(1/2)*(-12)^(1/3)*i)/2 - (-12)^(1/3)/2
```

The results indicate that *x* and *y* each have four roots. Two are real and two are complex. This points to the fact that MATLAB deals with complex numbers and, hence, not just real numbers. This is another important capability of MATLAB. In MATLAB by default (unless you reassign them) *i* and *j* represent the number known as $\sqrt{-1}$.

17.1.1 Polynomials

Everyone reading this book certainly knows about the quadratic equation. Most students of STEM subjects can recite, from memory, the roots of the quadratic equation; this is helpful to check the results found by applying MATLAB. The quadratic equation can be written as follows:

$$a x^2 + b x + c = 0. \tag{17.1}$$

Let us solve this equation using the `solve` utility in MATLAB. To do this we can execute the following script:

```
clear;clc
syms a b c x
solve(a*x^2 + b*x + c)
```

The answer in the Command Window is

```
ans =

  -(b + (b^2 - 4*a*c)^(1/2))/(2*a)
  -(b - (b^2 - 4*a*c)^(1/2))/(2*a)
```

Thus, as expected, the solutions of the quadratic equation are

$$x = -\frac{1}{2a}\left(b \pm \sqrt{b^2 - 4ac}\right) \tag{17.2}$$

For the two roots to be real numbers $b^2 \geq 4ac$. If the equality holds, then the two roots are equal. If $b^2 < 4ac$ the roots are complex and unequal. An alternative one-line command can be used to solve this equation; this can be done in the Command Window as follows

```
>> syms a b c x; solve(a*x^2+b*x+c)
```

The apostrophes indicate that what is between them is a symbolic expression. The output of both procedures of the application of `solve` is a 2 × 1 symbolic data-type array. Since this is a quadratic equation we expect to get two solutions.

Next, let us examine the cubic polynomial

$$a\,x^3 + b\,x^2 + c\,x + d = 0. \tag{17.3}$$

To find the roots of this equation we can execute the following script:

```
clear;clc
syms a b c d x
solve(a*x^3 + b*x^2 + c*x + d)
```

In this case, since we are solving a cubic, the answer (`ans`) is in a 3 × 1 symbolic data-type array. The answer is rather lengthy however and, hence, it is not reproduced here. It is produced in the command window.

Let us consider another example. Let us solve

$$ax^4 + cx = 0.$$

The following commands were executed in the Command Window:

```
>> syms a c x; solve(a*x^4 + c*x);
>> solution = simplify(ans);
>> solution

solution =

                              0
                     (-c/a)^(1/3)
   ((3^(1/2)*i - 1)*(-c/a)^(1/3))/2
  -((3^(1/2)*i + 1)*(-c/a)^(1/3))/2
```

```
>> latex(solution)

ans =

\left(\begin{array}{c} 0\\ {\left(-\frac{c}{a}\right)}^{\frac{1}{3}}\\
\frac{\left( - 1 + \sqrt{3}\, \mathrm{i}\right)\,
{\left(-\frac{c}{a}\right)}^{\frac{1}{3}}}{2}\\
-\frac{\left(1 + \sqrt{3}\, \mathrm{i}\right)\,
{\left(-\frac{c}{a}\right)}^{\frac{1}{3}}}{2} \end{array}\right)
```

This answer (`ans`) was used in the word processor used by the second author; the word processor implements Latex. This translation to Latex gives the following array of the four roots to this quartic equation:

$$
\begin{pmatrix}
0 \\
\left(-\frac{c}{a}\right)^{\frac{1}{3}} \\
\frac{\left(-1+\sqrt{3}\,\mathrm{i}\right)\left(-\frac{c}{a}\right)^{\frac{1}{3}}}{2} \\
-\frac{\left(1+\sqrt{3}\,\mathrm{i}\right)\left(-\frac{c}{a}\right)^{\frac{1}{3}}}{2}
\end{pmatrix}
$$

Note that two of the roots are complex numbers.

Finally, let us examine the solution of three simultaneous linear equations with the `solve` utility. Let us solve

$$x + y + z = 1, \quad 2x + 3y + z = 1, \quad x + y + 3z = 0.$$

Doing this in the Command Window, we get

```
>> syms x y z
>> [x y z] = solve(x + y + z - 1, 2*x + 3*y + z - 1, x + y + 3*z)
```

The solution of this system of equations leads to $x = 3$, $y = -3/2$ and $z = -1/2$.

Note that there are a variety of methods to implement the `solve` function (or solution procedure utility) within Symbolics. If numerical solutions are obtained, as in the last example, they need to be converted to *double* data types to use them in MATLAB scripts (this was already illustrated in Section 1.1.6).

17.1.2 Vectors

The single row or single column arrays are vectors. Let us consider the following 3-component vectors:

$$\mathbf{x}_1 = (a_1, b_1, c_1), \quad \mathbf{x}_2 = (a_2, b_2, c_2)$$

There are two ways to multiply these mathematical objects that most of the students are familiar. They are the *inner* or *dot* product and the *vector* or *cross* product. They can be found by applying the following script of Symbolic commands (from the editor).

```
format compact
syms a1 a2 b1 b2 c1 c2 real
x1 = [a1,b1,c1]
x2 = [a2,b2,c2]
x1_dot_x2 = dot(x1,x2)
x1_cross_x2 = cross(x1,x2)
format
```

The results in the Command Window are

```
x1 =
[ a1, b1, c1]
x2 =
[ a2, b2, c2]
x1_dot_x2 =
a1*a2 + b1*b2 + c1*c2
x1_cross_x2 =
[ b1*c2 - b2*c1, a2*c1 - a1*c2, a1*b2 - a2*b1]
```

The `format` compact was implemented to get the compact output illustrated above. The last command, viz., `format`, was implemented to reset the default output-display format in the Command Window. The solution can be written as follows:

$$\mathbf{x}_1 \cdot \mathbf{x}_2 = a_1 a_2 + b_1 b_2 + c_1 c_2$$

The dot product is a scalar.

$$\mathbf{x}_1 \times \mathbf{x}_2 = (b_1 c_2 - b_2 c_1, \ a_2 c_1 - a_1 c_2, \ a_1 b_2 - a_2 b_1)$$

The cross product is a vector. It is a three-component vector (the same number of components as the vectors that are multiplied to get this result).

17.1.3 Matrices

Let us consider the following mathematical objects known as matrices. We can add and subtract the two objects. We can multiply and divide the two objects.

We will do these operations using the Symbolic toolbox

$$\mathbf{M_a} = \begin{pmatrix} a_{11} & a_{12} \\ a_{21} & a_{22} \end{pmatrix}$$

$$\mathbf{M_b} = \begin{pmatrix} b_{11} & b_{12} \\ b_{21} & b_{22} \end{pmatrix}$$

The following script was executed in MATLAB:

```
format compact
syms a11 a12 a21 a22 b11 b12 b21 b22
Ma = [a11 a12; a21 a22]
Mb = [b11 b12; b21 b22]
Msum = Ma + Mb
Mproduct = Ma*Mb
```

The results in the Command Window are

```
Ma =
[ a11, a12]
[ a21, a22]
Mb =
[ b11, b12]
[ b21, b22]
Msum =
[ a11 + b11, a12 + b12]
[ a21 + b21, a22 + b22]
Mproduct =
[ a11*b11 + a12*b21, a11*b12 + a12*b22]
[ a21*b11 + a22*b21, a21*b12 + a22*b22]
```

The addition of the two matrices involve term-by-term addition.

$$\mathbf{M_a} \pm \mathbf{M_b} = \begin{pmatrix} a_{11} \pm b_{11} & a_{12} \pm b_{12} \\ a_{21} \pm b_{21} & a_{22} \pm b_{22} \end{pmatrix}.$$

The product leads to a matrix of the same size as the matrices that are multiplied; each term in the result is the multiplication of a row of $\mathbf{M_a}$ times a column on $\mathbf{M_b}$ summed together. Study the product matrix, as written below, to verify the matrix multiplication process. The product of the two matrices is

$$\mathbf{M_a} \, \mathbf{M_b} = \begin{pmatrix} a_{11} \, b_{11} + a_{12} \, b_{21} & a_{11} \, b_{12} + a_{12} \, b_{22} \\ a_{21} \, b_{11} + a_{22} \, b_{21} & a_{21} \, b_{12} + a_{22} \, b_{22} \end{pmatrix}.$$

Let us analyze a number of properties of the following matrix:

$$\mathbf{M_a} = \begin{pmatrix} c_{11} & c_{12} \\ c_{21} & c_{22} \end{pmatrix}$$

```
syms c11 c12 c21 c22 real
Mc = [c11 c12; c21 c22]
Mcdet = det(Mc);
Mcinv = inv(Mc);
Mc*inv(Mc);
disp(' Mc * Mcinv = Imatrix')
Imatrix = simple(ans)
Mcdet
Mcinv
[EigenVectorsMc EigenValuesMc] = eig(Mc)
```

Except for the eigenvectors and eigenvalues, the following output was in the Command Window. The latter were converted to Latex and are reported below.

```
Mc =
[ c11, c12]
[ c21, c22]
  Mc * Mcinv = Imatrix
Imatrix =
[ 1, 0]
[ 0, 1]
Mcdet =
c11*c22 - c12*c21
Mcinv =
[  c22/(c11*c22 - c12*c21), -c12/(c11*c22 - c12*c21)]
[ -c21/(c11*c22 - c12*c21),  c11/(c11*c22 - c12*c21)]
```

The eigenvalues associated with this matrix are

$$\begin{pmatrix} \frac{c_{11}}{2} + \frac{c_{22}}{2} - \frac{\sqrt{c_{11}{}^2 - 2c_{11}c_{22} + c_{22}{}^2 + 4c_{12}c_{21}}}{2} & 0 \\ 0 & \frac{c_{11}}{2} + \frac{c_{22}}{2} + \frac{\sqrt{c_{11}{}^2 - 2c_{11}c_{22} + c_{22}{}^2 + 4c_{12}c_{21}}}{2} \end{pmatrix}.$$

This is a result of diagonalizing the matrix (a process interpreted geometrically as a rotation of the coordinate system to find the coordinates in which the matrix only has finite diagonal terms). The coordinate system at which this occurs is given by the orthogonal eigenvectors. The corresponding eigenvectors for this example are

$$\begin{pmatrix} \frac{\frac{c_{11}}{2} + \frac{c_{22}}{2} - \frac{\sqrt{c_{11}{}^2 - 2c_{11}c_{22} + c_{22}{}^2 + 4c_{12}c_{21}}}{2}}{c_{21}} - \frac{c_{22}}{c_{21}} & \frac{\frac{c_{11}}{2} + \frac{c_{22}}{2} + \frac{\sqrt{c_{11}{}^2 - 2c_{11}c_{22} + c_{22}{}^2 + 4c_{12}c_{21}}}{2}}{c_{21}} - \frac{c_{22}}{c_{21}} \\ 1 & 1 \end{pmatrix}.$$

Let us next consider the analysis of a three-by-three symmetric matrix, viz.,

$$M = \begin{pmatrix} a & d & e \\ d & b & f \\ e & f & c \end{pmatrix}$$

This matrix was examined in detail by implementing the following script. It is an important mathematical object that plays an important role in, e.g., structural mechanics where it is the form of the stress tensor.

```
format compact
syms a b c d e f real
M = [a   d   e
     d   b   f
     e   f   c]
Mdet = det(M);
Minv = inv(M);
M*inv(M);
disp(' M * Minv = Imatrix')
Imatrix = simple(ans)
Mdet
Minv
[EigenVectorsM EigenValuesM] = eig(M)
format
```

The results in the Command Window, except for the rather lengthy eigenvalue, eigenvector and inverse expressions are as follows:

```
M =
[ a, d, e]
[ d, b, f]
[ e, f, c]
 M * Minv = Imatrix
Imatrix =
[ 1, 0, 0]
[ 0, 1, 0]
[ 0, 0, 1]
Mdet =
- c*d^2 + 2*d*e*f - b*e^2 - a*f^2 + a*b*c
```

We know that the inverse of M is determined correctly because $MM^{-1} = I$, where I is the identity matrix. It is left as an exercise to examine the inverse matrix and the results of the eigenvalue problem solved by the Symbolic tools in MATLAB.

If the matrix represents the state of stress at a point in a solid material, the eigenvalues would be the principal stresses and the eigenvectors the principal directions. This information is important in the analysis of the strength of materials. Similar matrices appear in modeling the mechanics of fluids.

17.2 CALCULUS

In this section we examine symbolic differentiation and integration. We will do this simultaneously because they are essentially inverse operation. Let us examine the quadratic function

$$f = ax^2 + bx + c. \tag{17.4}$$

The derivative of this equation is

$$\frac{df}{dx} = 2ax + b \tag{17.5}$$

This can be verified by applying the Symbolic tools in MATLAB. We can do this as follows:

```
format compact
syms x a b c real
f = a*x^2 + b*x + c
disp('dfdx = diff(f,x) was executed to get:')
dfdx = diff(f,x)
% Integration
disp('f = int(dfdx,x) was executed to get:')
f = int(dfdx,x)
disp(' Note that the constant of integration, c, is implied.')
disp(' Hence, to include it you need to add c to f:')
f = f + c
disp('f = expand(f) was executed to get:')
f = expand(f)
disp(' Thus, the function f is recovered (as expected).')
format
```

The output to the Command Window is

```
f =
a*x^2 + b*x + c
dfdx = diff(f,x) was executed to get:
dfdx =
b + 2*a*x
f = int(dfdx,x) was executed to get:
```

```
f =
x*(b + a*x)
  Note that the constant of integration, c, is implied.
  Hence, to include it you need to add c to f:
  f =
c + x*(b + a*x)
f = expand(f) was executed to get:
f =
a*x^2 + b*x + c
  Thus, the function f is recovered (as expected).
```

Thus, we verified the formula for df/dx. The next step is to verify the results of differentiation by integrating. This was done in the script given above by executing int(dfdx,x). The result obtained is f to within a constant. The integration tool does *not* explicitly include the arbitrary constant of integration. Hence, a constant needs to be added as illustrated in the above example. Then the function f with the added constant is expanded (expand(f)), leading to the result sought, i.e., the recovery of the original function f.

Let us examine a second example. It is the examination of

$$f = a \sin^2 x + b \cos x \tag{17.6}$$

Differentiating this formula and, subsequently, integrating df/dx, we get the following: The script in the editor that was executed is

```
format compact
syms x a b c real
f = a*sin(x)^2 + b*cos(x)
dfdx = diff(f,x)
% Integration
ff = int(dfdx,x)
ff = expand(ff)
format
```

The output to the Command Window is

```
f =
a*sin(x)^2 + b*cos(x)
dfdx =
2*a*cos(x)*sin(x) - b*sin(x)
ff =
2*(cos(x)/2 + 1/2)*(2*a + b - 2*a*(cos(x)/2 + 1/2))
ff =
a + b + b*cos(x) - a*cos(x)^2
```

Note that the result of the integration of df/dx is (it is `ff` in the script; it is f below)

$$f = a + b + b\cos x - a\cos^2 x$$

Substituting $\cos^2 x = 1 - \sin^2 x$, we get

$$f = a\sin^2 x + b\cos x + b + c$$

where c is the arbitrary constant of integration. Since it is arbitrary, let us select it as $c = -b$. Thus,

$$f = a\sin^2 x + b\cos x$$

This is the result sought; it is the same as given in the original equation, Eq. (17.6). Thus, these tools not only allow you to do differentiation and integration of a variety of problems you may confront in practice or wish to examine as part of your self-education, it helps provide insight into the connection between the operation of differentiation and the inverse operation of integration. We needed to be aware of the arbitrary constant of integration and what role it plays in the recovery of the formula that was differentiated. We needed to know that the derivative of a constant is identically zero.

17.3 LAPLACE AND Z TRANSFORMS

In this section the useful transforms known as the Laplace and the Z transforms are introduced by way of an example. Transforms are useful in applied mathematics as applied to examine engineering and scientific problems. There are other transforms available in Symbolics that could be investigated as well. We limit this primarily to the Laplace transform because it is introduced to undergraduates as part of the mathematics background necessary to examine dynamical systems. We examine the Z transform because it is interesting. In addition, we demonstrate the application of `ezplot` that is available in the Symbolic tools to help examine graphically various functions.

Let us examine the Laplace and the Z transforms of the function $y = \sin x$. Execute the following set of commands either in the Command Window or from the editor.

```
syms x t w
y = sin(x)
Ly = laplace(y)
yy = ilaplace(Ly,x)
Zy = ztrans(y)
ezplot(Ly)
hold on
```

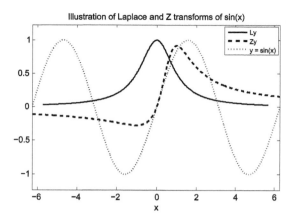

FIGURE 17.1
Comparison if $\sin x$, its Laplace and Z transforms.

```
ezplot(Zy)
ezplot(y)
```

The script in sequence determines the Laplace transform of y, Ly, it finds the inverse Laplace transform of the result to illustrate that you recover y, it determines the Z transform, Zy, (whatever that is) and it plots y, Ly and Zy. The results, as given in the Command Window, are as follows:

```
y =
sin(x)
Ly =
1/(s^2 + 1)
yy =
sin(x)
Zy =
(z*sin(1))/(z^2 - 2*cos(1)*z + 1)
```

The graphical results are in Fig. 17.1.

17.4 GENERALIZED FUNCTIONS*

In this section we examine very useful functions that are utilized in many scientific and engineering analyses. The first of the two functions examine is the Heaviside step function, $H(x)$. It is a function that is zero for $x < 0$, it is equal to 1/2 at $x = 0$ and it is equal to unity for $x > 1$. It is a built in function in MATLAB and in MATLAB Symbolics. To examine it the following script was executed.

FIGURE 17.2

The Heaviside step function, $H(x)$.

```
syms x
y = heaviside(x)
dydx = diff(y,x)
%
x = -1:.01:1;
y = heaviside(x);
plot(x,y,'--o'),title('The Heaviside step function')
xlabel('x');ylabel('y');
```

The output to the Command Window is

```
y =
heaviside(x)

dydx =
dirac(x)
```

Thus, the derivative of the Heaviside step function is the Dirac delta function (which we will examine next). A graphical illustration of the Heaviside step function is given in Fig. 17.2.

To use the Symbolic tools to examine the Dirac delta function let us execute the following script:

```
syms x
y = dirac(x)
I1 = int(y,x)
I2 = int(y,x,-1,1)
```

The results in the Command Window are as follows:

```
y =
dirac(x)
I1 =
heaviside(x)
I2 =
1
```

This illustrates that the indefinite integral of the Dirac function is the Heaviside function. It also illustrates that the definite integral that includes $x = 0$ is one. The Dirac function is infinite at $x = o$ and it is zero everywhere else. It $x = 0$ is not included in the range of integration the integral is zero. Note that where the delta function is infinite and, hence, where the Heaviside step function jumps from zero to one can be changed by changing the x coordinate to another origin where the argument of the two functions is equal to zero.

17.5 DIFFERENTIAL EQUATIONS

The function dsolve is an ordinary differential equation and a system of ordinary differential equation (ODE) solver. We will examine two examples in this section. One is the solution of a single ODE and a system of ODEs. Let us examine the equation

$$\frac{d^3 y}{dx^3} = a x^2.$$

To solve this equation all we need to do is to execute the following command:

```
>> dsolve('D3y= a*x^2','x')
```

The result of executing this command is

```
ans =
(a*x^5)/60 + (C2*x^2)/2 + C3*x + C4
```

Thus, the solution is

$$\frac{a x^5}{60} + \frac{C_1 x^2}{2} + C_2 x + C_3$$

where C_1, C_2 and C_3 are the three arbitrary constants of integration. The procedure for solving a differential equation is to integrate. In this case because it is a third-order differential equation that we solved, it required three integrations and, hence, the result is to within three arbitrary constants.

In the next example we solve two equations simultaneously. In addition, we specify constraints (or boundary conditions). The equations to be solved are

$$\frac{df}{dt} = 3f + 4g, \quad \frac{dg}{dt} = -4f + 3g.$$

This system of equations, since it is second order, is subject to two constraints. In this example they are $f(0) = 0$ and $g(0) - 1$, which can be interpreted as initial conditions. The solution can be obtained by implementing the following command in MATLAB:

```
[f, g] = dsolve('Df = 3*f + 4*g, Dg = -4*f + 3*g','f(0) = 0, g(0) = 1')
```

The solution is

```
f =
sin(4*t)*exp(3*t)
g =
cos(4*t)*exp(3*t)
```

Thus,

$$f = \sin(4t)\,\exp^{3t}, \quad g = \cos(4t)\,\exp^{3t}.$$

The `dsolve` function is a powerful function to apply to try to find analytic solutions to systems of ODEs.

17.6 IMPLEMENTATION OF FUNTOOL, MUPAD AND HELP

This is the last section of this chapter. In this section we illustrate `funtool`, an interesting utility to examine graphically a variety of functions. The MuPAD capability to create your own notebooks associated with your own investigations is introduced. Finally, the help documents that come with MATLAB that are related to Symbolics is reviewed.

17.6.1 The funtool

In the Command Window type `funtool` and execute the command. This is done as follows:

```
>> funtool
```

FIGURE 17.3
The function tool.

The three windows illustrated in Fig. 17.3 are opened. If you type a function you are interested in examine in the space defining f, you examine it by hitting **<Enter>**. If you are interested in examining the capabilities of this tool click the Demo button. This tool provides an interesting and convenient way to examine graphically any function you wish to investigate.

17.6.2 The MuPAD notebook* and Symbolic help

If you are interested in creating a notebook within MATLAB to examine a subject under investigation that requires symbolic among other tools to study, then the MuPAD notebook environment may be of interest. To open a new notebook type the following command in the Command Window and execute it.

```
>> mupad
```

The window illustrated in Fig. 17.4 is opened. Examine the tools on the tool bar to learn how to input text as well as enter commands that allow you to execute Symbolic commands within this notebook environment.

Finally, to get help on using mupad and other features of the Symbolic tools click the circled question mark, '?', just below the word "Help" on the top tool bar. This opened the help window illustrated in Fig. 17.5. To get to the Symbolics help click on Symbolic Math Toolbox in the contents palette. The information in the right-hand pane will change to the information in Fig. 17.6. Note that there are examples, help for MuPAD and demos that are quite useful.

FIGURE 17.4
The notebook tool.

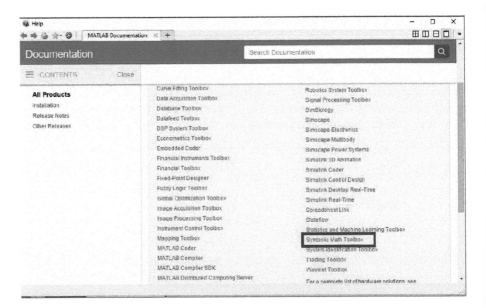

FIGURE 17.5
The MATLAB help documents: Click item in red box on the list of tools.

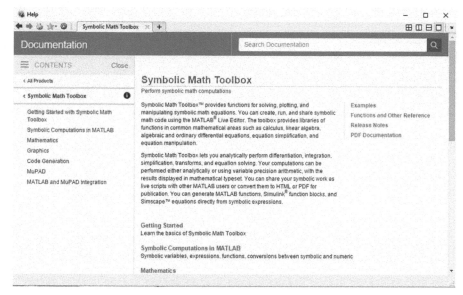

FIGURE 17.6
The help documents on the Symbolic tools.

EXERCISES

17.1 Find the derivatives of $\cos x$, and $x^2 \exp^{2x}$.

17.2 Integrate the function $y = 3x^2$ from $x = 0$ to $x = 10$.

Syntax: Quick Reference

In this appendix we offer examples of the MATLAB syntax most commonly used in this book.

A.1 EXPRESSIONS

```
x = 2 ^ (2 * 3) / 4;
x = A \ b;              % solution of linear equations
a == 0 & b < 0          % a equals 0 AND b less than 0
a ~= 4 | b > 0          % a not equal to 4 OR b greater than 0
```

A.2 FUNCTION M-FILES

```
function y=f(x)                          % save as f.m
% comment for help

function [out1, out2] = plonk(in1, in2, in3)  % save as plonk.m
% Three input arguments, two outputs
  ...

function junk      % no input/output arguments; save as junk.m

[t, x] = ode45(@lorenz, [0 10], x0);  % function handle with @
```

A.3 GRAPHICS

```
plot(x, y), grid     % plots vector y against vector x on a grid

plot(x, y, 'b--')    % plots a blue dashed line

plot(x, y, 'go')     % plots green circles

plot(y)  % if y is a vector plots elements against row numbers
         % if y is a matrix, plots columns against row numbers
```

379

```
plot(x1, y1, x2, y2) % plots y1 against x1 and
                        y2 against x2 on same graph

semilogy(x, y)    % uses a log10 scale for y

polar(theta, r)   % generates a polar plot
```

A.4 if **AND** switch

```
if condition
  statement        % executed if condition true
end;

if condition
  statement1       % executed if condition true
else
  statement2       % executed if condition false
end;

if a == 0         % test for equality
  x = -c / b;
else
  x = -b / (2*a);
end;

if condition1     % jumps off ladder at first true condition
  statement1
elseif condition2 % elseif one word!
  statement2
elseif condition3
  statement3
...
else
  statementE
end;
if condition statement1, else statement2, end   % command line

switch lower(expr)              % expr is string or scalar
  case {'linear','bilinear'}
    disp('Method is linear')
  case 'cubic'
    disp('Method is cubic')
  case 'nearest'
    disp('Method is nearest')
```

```
  otherwise
    disp('Unknown method.')
end
```

A.5 for **AND** while

```
for i = 1:n          % repeats statements n times
   statements
end;

for i = 1:3:8        % i takes values 1, 4, 7
   ...
end;

for i = 5:-2:0       % i takes values 5, 3, 1
   ...
end;

for i = v            % index i takes on each element of vector v
   statements
end;

for v = a            % index v takes on each column of matrix a
   statements
end;

for i = 1:n, statements, end      % command line version

try,
   statements,
catch,
   statements,
end

while condition    % repeats statements while condition is true
   statements
end;

while condition statements, end    % command line version
```

A.6 INPUT/OUTPUT

```
disp( x )
```

```
disp( 'Hello there' )

disp([a b])        % two scalars on one line

disp([x' y'])      % two columns (vectors x and y
                                  must be same length)

disp( ['The answer is ', num2str(x)] )

fprintf( '\n' )                    %  new line

fprintf( '%5.1f\n', 1.23 )        %  **1.2

fprintf( '%12.2e\n', 0.123 )      %  ***1.23e-001

fprintf( '%4.0f and %7.2f\n', 12.34, -5.6789 )
                                  %  **12 and **-5.68

fprintf( 'Answers are: %g %g\n', x, y )
                                  % matlab decides on format

fprintf( '%10s\n', str )          % left-justified string

x = input( 'Enter value of x: ' )

name = input( 'Enter your name without apostrophes: ', 's' )
```

A.7 load/save

```
load filename          % retrieves all variables
                           from binary file filename.mat
load x.dat             % imports matrix x from ASCII file x.dat
save filename x y z     % saves x y and z in filename.mat
save                   % saves all workspace variables
                           in matlab.mat
save filename x /ascii  % saves x in filename (as ASCII file)
```

A.8 VECTORS AND MATRICES

```
a(3,:)                 % third row

a(:,2)                 % second column
```

```
v(1:2:9)                % every second element from 1 to 9
v([2 4 5]) = [ ]        % removes second, fourth and fifth elements

v(logical([0 1 0 1 0])) % second and fourth elements only

v'                      % transpose
```

Operators

Precedence	Operators
1	()
2	^ .^ ' .' (pure transpose)
3	+ (unary plus) - (unary minus) ~ (NOT)
4	* / \ .* ./ .\
5	+ (addition) - (subtraction)
6	:
7	> < >= <= == ~=
8	& (AND)
9	\| (OR)
*(See **Help** on operator precedence.)*	

Command and Function: Quick Reference

This appendix lists most of the MATLAB commands and functions used in the text, along with a few others. For a complete list by category (with links to detailed descriptions) see **MATLAB: Reference: MATLAB Function Reference: Functions by Category** in the online documentation.

The command `help` by itself displays a list of all the function categories (each in its own directory).

C.1 GENERAL-PURPOSE COMMANDS

demo	Run demos
help	Online help
helpwin	Display categories of functions with links to each category
lookfor	Keyword search through `help` entries
type	List M-files
what	Directory listing of M- and MAT-files
which	Locate functions and files

C.1.1 Managing variables and the workspace

clear	Clear variables and functions from memory
disp	Display matrix or text
length	Length of vector
load	Retrieve variables from disk
save	Save workspace variables to disk
size	Array dimensions
who, whos	List variables in workspace

C.1.2 Files and the operating system

beep	Produce beep sound
cd	Change current working directory
delete	Delete file

387

diary	Save text of MATLAB session
dir	Directory listing
edit	Edit M-file
!	Execute operating system command

C.1.3 Controlling the Command Window

clc	Clear Command Window
echo	Echo commands in script
format	Set output format for disp
home	Send cursor home
more	Control paged output

C.1.4 Starting and quitting MATLAB

exit	Terminate MATLAB
quit	Terminate MATLAB
startup	Execute M-file when MATLAB starts

C.2 LOGICAL FUNCTIONS

all	True if all elements of vector are true (nonzero)
any	True if any element of vector is true
exist	Check if variable or file exists
find	Find indices of nonzero elements
is*	Detect various states
logical	Convert numeric values to logical values

C.3 MATLAB PROGRAMMING TOOLS

error	Display error message
eval	Interpret string containing MATLAB expression
feval	Function evaluation
for	Repeat statement specific number of times
global	Define global variable
if	Conditionally execute statements
persistent	Define persistent variable
switch	Switch among several cases
try	Begin try block
while	Repeat statements conditionally

C.3.1 Interactive input

input	Prompt user for input
keyboard	Invoke keyboard as script file
menu	Generate menu of choices for user input
pause	Wait for user response

C.4 MATRICES

eye	Identity matrix
linspace	Vector with linearly spaced elements
ones	Matrix of ones
rand	Uniformly distributed random numbers and arrays
randn	Normally distributed random numbers and arrays
zeros	Matrix of zeros
: (colon)	Vector with regularly spaced elements

C.4.1 Special variables and constants

ans	Most recent answer
eps	Floating-point relative accuracy
i or j	$\sqrt{-1}$
Inf	Infinity
NaN	Not-a-Number
nargin, nargout	Number of actual function arguments
pi	$3.14159\,26535\,897\ldots$
realmax	Largest positive floating-point number
realmin	Smallest positive floating-point number
varargin, varargout	Pass or return variable numbers of arguments

C.4.2 Time and date

calendar	Calendar
clock	Wall clock (complete date and time)
date	Actual date
etime	Elapsed time
tic, toc	Stopwatch
weekday	Day of week

C.4.3 Matrix manipulation

cat	Concatenate arrays
diag	Create or extract diagonal

fliplr	Flip in left/right direction
flipud	Flip in up/down direction
repmat	Replicate and tile array
reshape	Change shape
rot90	Rotate 90°
tril	Extract lower tridiagonal part
triu	Extract upper tridiagonal part

C.4.4 Specialized matrices

gallery	Test matrices
hilb	Hilbert matrix
magic	Magic square
pascal	Pascal matrix
wilkinson	Wilkinson's eigenvalue test matrix

C.5 MATHEMATICAL FUNCTIONS

abs	Absolute value
acos, acosh	Inverse cosine, inverse hyperbolic cosine
acot, acoth	Inverse cotangent, inverse hyperbolic cotangent
acsc, acsch	Inverse cosecant, and inverse hyperbolic cosecant
angle	Phase angle
asec, asech	Inverse secant, inverse hyperbolic secant
asin, asinh	Inverse sine, inverse hyperbolic sine
atan, atanh	Inverse tangent (two quadrant), inverse hyperbolic tangent
atan2	Inverse tangent (four quadrant)
bessel	Bessel function
ceil	Round up
conj	Complex conjugate
cos, cosh	Cosine, hyperbolic cosine
cot, coth	Cotangent, hyperbolic cotangent
csc, csch	Cosecant, hyperbolic cosecant
erf	Error function
exp	Exponential
fix	Round toward zero
floor	Round down
gamma	Gamma function
imag	Imaginary part
log	Natural logarithm
log2	Dissect floating-point numbers into exponent and mantissa

log10	Common logarithm
mod	Modulus (signed remainder after division)
rat	Rational approximation
real	Real part
rem	Remainder after division
round	Round toward nearest integer
sec, sech	Secant, hyperbolic secant
sign	Signum function
sin, sinh	Sine, hyperbolic sine
sqrt	Square root
tan, tanh	Tangent, hyperbolic tangent

C.6 MATRIX FUNCTIONS

det	Determinant
eig	Eigenvalues and eigenvectors
expm	Matrix exponential
inv	Matrix inverse
poly	Characteristic polynomial
rank	Number of linearly independent rows or columns
rcond	Condition estimator
trace	Sum of diagonal elements
{}\ and /	Linear equation solution

C.7 DATA ANALYSIS

cumprod	Cumulative product
cumsum	Cumulative sum
diff	Difference function
fft	One-dimensional fast Fourier transform
max	Largest element
mean	Average value of elements
median	Median value of elements
min	Smallest element
prod	Product of elements
sort	Sort in ascending order
std	Standard deviation
sum	Sum of elements
trapz	Trapezoidal rule for numerical integration

C.8 POLYNOMIAL FUNCTIONS

polyfit	Fit polynomial to data
polyval	Evaluate polynomial
roots	Find polynomial roots

C.9 FUNCTION FUNCTIONS

bvp4c	Solve two-point boundary value problems for ODEs
fmin	Minimize function of one variable
fmins	Minimize function of several variables
fzero	Find zero of function of one variable
ode23, ode23s, ode45	Solve initial value problems for ODEs
quad	Numerical integration

C.10 SPARSE MATRIX FUNCTIONS

full	Convert sparse matrix to full matrix
sparse	Construct sparse matrix from nonzeros and subscripts
spy	Visualize sparse matrix

C.11 CHARACTER STRING FUNCTIONS

char	Characters from ASCII codes
double	ASCII codes of characters
lower	Convert string to lower case
sprintf	Write formatted data to string
str2mat	String-to-matrix conversion
strcat	String concatenation
strcmp	Compare strings
upper	Convert string to upper case

C.12 FILE I/O FUNCTIONS

fclose	Close one or more open files
feof	Test for end-of-file
fopen	Open file or obtain information about open files
fprintf	Write formatted data to file
fread	Read binary data from file

fscanf	Read formatted data from file
fseek	Set file position indicator
ftell	Get file position indicator
fwrite	Write binary data to file

C.13 2D GRAPHICS

bar	Bar graph
grid	Grid lines
hist	Histogram plot
loglog	Log-log scale plot
plot	Linear plot
polar	Polar coordinate plot
semilogx	Semi-log scale plot
semilogy	Semi-log scale plot
text	Text annotation
title	Graph title
xlabel	x-axis label
ylabel	y-axis label
zoom	Zoom in and out on 2D plot

C.14 3D GRAPHICS

clabel	Contour plot elevation label
comet3	Animated 3D plot
contour	Contour plot in 2D
contour3	Contour plot in 3D
mesh	3D mesh surface
meshc	3D mesh surface with contour plot
meshgrid	X and Y arrays for 3D plots
plot3	Line plot in 3D
quiver	Quiver plot
surf	Shaded surface
surfl	Shaded surface with lighting
view	Rotation of 3D figure
zlabel	z-axis label

C.15 GENERAL

axes	Create axes object
axis	Control axis scaling and appearance

cla	Clear axes
clf	Clear current figure
colorbar	Display color bar (color scale)
colormap	Set color look-up table (list of colormaps)
drawnow	Complete any pending drawing
figure	Create figure (graph) windows
fplot	Plot function
gca	Get current axes handle
gcf	Get current figure handle
gco	Return handle of current graphics object
get	Get graphics object properties
ginput	Graphical input from mouse or cursor
gtext	Mouse placement of text
set	Set graphics object properties
subplot	Create axes in tiled positions

Solutions to Selected Exercises

CHAPTER 1

1.1
```
a = 3;
b = 5;
sum        = a + b;
difference = a - b;
product    = a * b;
quotient   = a / b;
```

CHAPTER 2

2.1 (a) Comma should be replaced by decimal point
 (e) Asterisk should be omitted
 (f) Exponent must be integer
 (h) Comma should be replaced by decimal point

2.2 (b) Decimal point not allowed
 (c) First character must be letter
 (d) Quotes not allowed
 (h) Blanks not allowed
 (i) Allowed but not recommended
 (k) Asterisk not allowed
 (l) Allowed but not recommended

2.3 (a) `p + w/u`
 (b) `p + w/(u + v)`
 (c) `(p + w/(u+v))/(p + w/(u-v))`
 (d) `sqrt(x)`
 (e) `y^(y+z)`
 (f) `x^(y^z)`
 (g) `(x^y)^z`
 (h) `x - x^3/(3*2) + x^5/(5*4*3*2)`

2.4 (a) `i = i + 1`
 (b) `i = i^3 + j`
 (c)
```
if e > f
   g = e
else
```

```
              g = f
          end
    (d)   if d > 0
              x = -b
          end
    (e)   x = (a + b)/(c * d)
2.5   (a)  Expression not allowed on left-hand side
      (b)  Left-hand side must be valid variable name
      (c)  Left-hand side must be valid variable name
2.6   a = 2;
      b = -10;
      c = 12;
      x = (-b + sqrt(b ^ 2 - 4 * a * c)) / (2 * a)
2.7   gallons = input('Enter gallons: ');
      pints = input('Enter pints: ');
      pints = pints + 8 * gallons;
      liters = pints / 1.76
2.8   distance = 528;
      liters = 46.23;
      kml = distance / liters;
      l100km = 100 / kml;
      disp( 'Distance    Liters used    km/L    L/100km' );
      disp( [distance liters kml l100km] );
2.9   t = a;
      a = b;
      b = t;
2.10  a = [a b];      % make 'a' into a vector
      b = a(1);
      a(1) = [];
2.11   (a)  c = input('Enter Celsius temperature: ');
            f = 9 * c / 5 + 32;
            disp( ['The Fahrenheit temperature is:' num2str(f)] );
       (b)  c = 20 : 30;
            f = 9 * c / 5 + 32;
            format bank;
            disp('        Celsius      Fahrenheit');
            disp([c'      f']);
2.12  degrees = 0 : 10 : 360;
      radians = degrees / 180 * pi;
      format bank;
      disp('        Degrees      Radians');
      disp([degrees'      radians']);
2.13  degrees = 0 : 30 : 360;
      radians = degrees / 180 * pi;
      sines = sin(radians);
```

```
        cosines = cos(radians);
        tans = tan(radians);
        table = [degrees' sines' cosines' tans']
2.14    for int = 10 : 20
          disp( [int sqrt(int)] );
        end
2.15    sum(2 : 2 : 200)
2.16    m = [5  8  0  10  3  8  5  7  9  4];
        disp( mean(m) )
2.17    x = 2.0833,  a = 4
2.18    % With for loop
        i = 1;
        x = 0;
        for a = i : i : 4
          x = x + i / a;
        end

        % With vectors
        i = 1;
        a = i : i : 4;
        x = i ./ a;
        sum(x)
2.19  (b)  n = input('Number of terms? ');
           k = 1 : n;
           s = 1 ./ (k .\char136 2);
           disp(sqrt(6 * sum(s)))
2.21    r = 5;
        c = 10;
        l = 4;
        e = 2;
        w = 2;
        i = e / sqrt(r ^ 2 + (2 * pi * w * l -
                 1 / (2 * pi * w * c)) ^ 2)
2.22    con = [200 500 700 1000 1500];
        for units = con
          if units <= 500
            cost = 0.02 * units;
          elseif units <= 1000
            cost = 10 + 0.05 * (units - 500);
          else
            cost = 35 + 0.1 * (units - 1000);
          end
          charge = 5 + cost;
          disp( charge )
        end
```

```
2.24  money = 1000;
      for month = 1 : 12
        money = money * 1.01;
      end
2.26  t = 1790 : 10 : 2000;
      p = 197273000 ./ (1 + exp(-0.03134 * (t - 1913.25)));
      disp([t'   p']);
      pause;
      plot(t,p);
2.27  (a)  r = 0.15;
           l = 50000;
           n = 20;

           p = r * l * (1 + r / 12) ^ (12 * n) / ...
               (12 * ((1 + r / 12) ^ (12 * n)  1))
2.28  (a)  r = 0.15;
           l = 50000;
           p = 800;
           n = log(p / (p - r * l / 12)) / (12 * log(1 + r / 12))
```

CHAPTER 3

3.1 You should get a picture of tangents to a curve.

3.2 (a) 4

　　　(b) 2

　　　(c) Algorithm (attributed to Euclid) finds the HCF (highest common factor) of two numbers using the fact that it divides exactly into the difference between the two numbers, and that, if the numbers are equal, they are equal to their HCF.

```
3.3  f = input('Enter Fahrenheit temperature: ');
     c = 5 / 9 * (f {\minuscda} 32);
     disp( ['The Celsius temperature is: ' num2str(c)] );
3.4  a = input('Enter first number: ');
     b = input('Enter second number: ');
     if a < b
       disp( [ num2str(b) ' is larger.'] );
     elseif a > b
       disp( [ num2str(a) ' is larger.'] );
     else
       disp( 'Numbers are equal.' );
     end
```

3.6 1. Input a, b, c, d, e, f
 2. $u = ae - db$, $v = ec - bf$
 3. If $u = 0$ and $v = 0$, then
 Lines coincide
 Otherwise if $u = 0$ and $v \neq 0$, then
 Lines are parallel
 Otherwise
 $x = v/u$, $y = (af - dc)/u$
 Print x, y
 4. Stop

```
a = input('Enter a: ');
b = input('Enter b: ');
c = input('Enter c: ');
d = input('Enter d: ');
e = input('Enter e: ');
f = input('Enter f: ');
u = a * e - b * d;
v = c * e - b * f;
if u == 0
  if v == 0
    disp('Lines coincide.');
  else
    disp('Lines are parallel.');
  end
else
  x = v / u;
  y = (a * f - d * c) / u;
  disp( [x y] );
end
```

CHAPTER 4

4.2 (a) `log(x + x ^ 2 + a ^ 2)`
 (b) `(exp(3 * t) + t ^ 2 * sin(4 * t)) * (cos(3 * t)) ^ 2`
 (c) `4 * atan(1)`
 (d) `sec(x)^2 + cot(x)`
 (e) `atan(a / x)`

4.3
```
m = input('Enter length in meters: ');
inches = m * 39.37;
feet = fix(inches / 12);
inches = rem(inches, 12);
yards = fix(feet / 3);
feet = rem(feet, 3);
disp( [yards feet inches] );
```

```
4.5   a = 10;
      x = 1;
      k = input('How many terms do you want? ');
      for n = 1 : k
        x = a * x / n;
        if rem(n, 10) == 0
          disp( [n x] );
        end
      end
4.6   secs = input('Enter seconds: ');
      mins = fix(secs / 60);
      secs = rem(secs, 60);
      hours = fix(mins / 60);
      mins = rem(mins, 60);
      disp( [hours mins secs] );
```

CHAPTER 5

5.2 (a) 1 1 0
 (b) 0 1 0
 (c) 1 0 1
 (d) 0 1 1
 (e) 1 1 1
 (f) 0 0 0
 (g) 0 2
 (h) 0 0 1

```
5.3   neg = sum(x < 0);
      pos = sum(x > 0);
      zero = sum(x == 0);
5.7   units = [200 500 700 1000 1500];
      cost = 10 * (units > 500) + 25 * (units > 1000) + 5;
      cost = cost + 0.02 * (units <= 500) .* units;
      cost = cost + 0.05 * (units > 500 & units <= 1000) .*
                          (units - 500);
      cost = cost + 0.1 * (units > 1000) .* (units - 1000);
```

CHAPTER 6

```
6.6   function x = mygauss(a, b)
      n = length(a);

      a(:,n+1) = b;

      for k = 1:n
        a(k,:) = a(k,:)/a(k,k);     % pivot element must be 1
```

```
      for i = 1:n
        if i ~= k
          a(i,:) = a(i,:) - a(i,k) * a(k,:);
        end
      end

   end

   % solution is in column n+1 of a:
   x = a(:,n+1);
```

CHAPTER 7

```
7.1  function pretty(n, ch)
     line = char(double(ch)*ones(1,n));
     disp(line)
7.2  function newquot(fn)
     x = 1;
     h = 1;
     for i = 1 : 10
       df = (feval(fn, x + h) - feval(fn, x)) / h;
       disp( [h, df] );
       h = h / 10;
     end
7.3  function y = double(x)
     y = x * 2;
7.4  function [xout, yout] = swop(x, y)
     xout = y;
     yout = x;
7.6  % Script file
     for i = 0 : 0.1 : 4
       disp( [i, phi(i)] );
     end

     % Function file phi.m
     function y = phi(x)
     a = 0.4361836;
     b = -0.1201676;
     c = 0.937298;
     r = exp(-0.5 * x * x) / sqrt(2 * pi);
     t = 1 / (1 + 0.3326 * x);
     y = 0.5 - r * (a * t + b * t * t + c * t ^ 3);
7.8  function y = f(n)
     if n > 1
```

```
      y = f(n - 1) + f(n - 2);
   else
      y = 1;
   end
```

CHAPTER 8

8.1
```
balance = 1000;
for years = 1 : 10
  for months = 1 : 12
    balance = balance * 1.01;
  end
  disp( [years balance] );
end
```

8.2 (a)
```
terms = 100;
pi = 0;
sign = 1;
for n = 1 : terms
  pi = pi + sign * 4 / (2 * n - 1);
  sign = sign * (-1);
end
```

(b)
```
terms = 100;
pi = 0;
for n = 1 : terms
  pi = pi + 8 / ((4 * n - 3) * (4 * n - 1));
end
```

8.3
```
a = 1;
n = 6;
for i = 1 : 10
  n = 2 * n;
  a = sqrt(2 - sqrt(4 - a * a));
  l = n * a / 2;
  u = l / sqrt(1 - a * a / 2);
  p = (u + l) / 2;
  e = (u - l) / 2;
  disp( [n, p, e] );
end
```

8.5
```
x = 0.1;
for i = 1 : 7
  e = (1 + x) ^ (1 / x);
  disp( [x, e] );
  x = x / 10;
end
```

8.6
```
n = 6;
T = 1;
i = 0;
for t = 0:0.1:1
  i = i + 1;
  F(i) = 0;
  for k = 0 : n
    F(i) = F(i) + 1 / (2 * k + 1) * sin((2 * k + 1) *
                                    pi * t / T);
  end
  F(i) = F(i) * 4 / pi;
end
t = 0:0.1:1;
disp( [t' F'] )
plot(t, F)
```

8.8
```
sum = 0;
terms = 0;
while (sum + terms) <= 100
  terms = terms + 1;
  sum = sum + terms;
end
disp( [terms, sum] );
```

8.10
```
m = 44;
n = 28;
while m ~= n
  while m > n
    m = m - n;
  end
  while n > m
    n = n - m;
  end
end
disp(m);
```

CHAPTER 9

9.1
```
t = 1790:2000;
P = 197273000 ./ (1+exp(-0.03134*(t-1913.25)));
plot(t, P), hold, xlabel('Year'), ylabel('Population size')
census = [3929 5308 7240 9638 12866 17069 23192 31443 38558
          ...
          50156 62948 75995 91972 105711 122775
          131669 150697];
census = 1000 * census;
plot(1790:10:1950, census, 'o'), hold off
```

9.2
```
a = 2;
q = 1.25;
th = 0:pi/40:5*pi;
subplot(2,2,1)
plot(a*th.*cos(th), a*th.*sin(th)), ...
    title('(a) Archimedes')    % or use polar
subplot(2,2,2)
plot(a/2*q.^th.*cos(th), a/2*q.^th.*sin(th)), ...
    title('(b) Logarithmic')   % or use polar
```

9.4
```
n=1:1000;
d = 137.51;
th = pi*d*n/180;
r = sqrt(n);
plot(r.*cos(th), r.*sin(th), 'o')
```

9.6
```
y(1) = 0.2;
r = 3.738;
for k = 1:600
  y(k+1) = r*y(k)*(1 - y(k));
end
plot(y, '.w')
```

CHAPTER 11

11.1
```
x = 2;
h = 10;
for i = 1 : 20
  h = h / 10;
  dx = ((x + h) ^ 2 - x * x) / h;
  disp( [h, dx] );
end
```

CHAPTER 13

13.1
```
heads = rand(1, 50) < 0.5;
tails = ~heads;
heads = heads * double('H');
tails = tails * double('T');
coins = char(heads + tails)
```

13.2
```
bingo = 1 : 99;
for i = 1 : 99
  temp = bingo(i);
  swop = floor(rand * 99 + 1);
  bingo(i) = bingo(swop);
  bingo(swop) = temp;
end
```

```
      for i = 1 : 10 : 81
        disp(bingo(i : i + 9))
      end
      disp(bingo(91 : 99))
13.4  circle = 0;
      square = 1000;
      for i = 1 : square
        x = 2 * rand - 1;
        y = 2 * rand - 1;
        if (x * x + y * y) < 1
          circle = circle + 1;
        end
      end
      disp( circle / square * 4 );
```

CHAPTER 14

14.1 (a) Real roots at 1.856 and −1.697; complex roots at −0.0791 ± 1.780i
 (b) 0.589, 3.096, 6.285, ... (roots get closer to multiples of π)
 (c) 1, 2, 5
 (d) 1.303
 (e) −3.997, 4.988, 2.241, 1.768

14.2 Successive bisections: 1.5, 1.25, 1.375, 1.4375, and 1.40625 (exact answer: 1.414214 ..., so the last bisection is within the required error)

14.3 22 (exact answer: 21.3333)

14.4 After 30 years, exact answer: 2 117 ($1000\,e^{rt}$)

14.6 The differential equations to be solved are

$$\mathrm{d}S/\mathrm{d}t = -r_1 S,$$
$$\mathrm{d}Y/\mathrm{d}t = r_1 S - r_2 Y.$$

The exact solution after 8 hours is $S = 6.450 \times 10^{25}$ and $Y = 2.312 \times 10^{26}$.

14.8
```
function s = simp(f, a, b, h)
x1 = a + 2 * h : 2 * h : b {\minuscda} 2 * h;
sum1 = sum(feval(f, x1));
x2 = a + h : 2 * h : b {\minuscda} h;
sum2 = sum(feval(f, x2));
s = h / 3 * (feval(f, a) + feval(f, b) +
            2 * sum1 + 4 * sum2);
```
With 10 intervals ($n = 5$), luminous efficiency is 14.512725%. With 20 intervals, it is 14.512667%. These results justify the use of 10 intervals in any further computations. This is a standard way to test the accuracy of a numerical method: halve the step-length and see how much the solution changes.

14.9
```
% Command Window
beta = 1;
ep = 0.5;
[t, x] = ode45(@vdpol, [0 20], [0; 1], [], beta, ep);
plot(x(:,1), x(:,2))

% Function file vdpol.m
function f = vdpol(t, x, b, ep)
f = zeros(2,1);
f(1) = x(2);
f(2) = ep * (1 - x(1)^2) * x(2) - b^2 * x(1);
```

Index

Symbols

\, 44, 153
%, 42
@
 function handle creation, 171
&, 115
', 116, 130
.', 116
.*, 46
./, 46
.\, 46
.^, 46
:, 9, 130
;, 7, 48
~, 115
|, 115
π, 8, 193
 estimation of, 294
3-D plot, 206

A

abs, 72, 100
acos, 100
acosh, 100
air resistance, 271
algebra, 360
algorithm, 83
all, 119, 137
 example, 120
AND, 115
angle, 72
animation, 222
 comet3, 207
ans, 33, 47
any, 119
 example, 120
apostrophe, 130

argument
 input, 93
 input, pass by reference, 169
 input, pass by value, 169
arithmetic operator, 44
array, 9, 32
 2-D, 127
 cell, 251
 multi-dimensional, 142
Array Editor, 33
array operation, 46, 138
asin, 100
asinh, 100
aspect ratio, 72
assignment, 7
assignment statement, 48
atan, 100
atan2, 72, 100
atanh, 100
axis, 200
 equal, 72
 normal, 72
axis limits, 200
azimuth, 214

B

bacteria division, 288
bacteria growth, 294, 309
bar graph, 245
base workspace, 167
best fit, 155
binomial coefficient, 181
Bisection method, 301
bit, 33
bitwise logical operations, 115
break, 184, 190
breakpoint, 175

Bubble Sort, 246
bug, 257
byte, 33

C

calculus, 360
calendar, 8
camel caps, 32
cantilever beam, 265
cartesian co-ordinates, 204
case, 71
case sensitivity, 32
ceil, 100
cell, 251, 252
cell array, 251
celldisp, 254
cellplot, 254
chaos, 224, 234, 314
 butterfly, 317
 trajectory, 317
cla, 203
clc, 8
clear, 33
clear M-file, 168
clf, 203
clipboard, 25
clock, 61, 100
colon operator, 9, 35, 130
colon operator, initialization, 35
colon operator, precedence, 45
color, 200
colorbar, 227
colormap, 225
comet3, 207
command, 49
command line, 6
command line editing, 6
Command Window, 5

command/function duality, 173
comment, 42
compiler, 28
complex conjugate transpose, 72
complex numbers, 71
condition estimator, 155
conj, 72
Contents.m, 167
context menu, 28
continuation of statement, 42, 48
continue, 190, 191
contour, 209
contour plot, 22
contour3, 210
cos, 101
cosh, 101
cot, 101
Crank-Nicolson method, 319
csc, 101
cumsum, 101
current directory, 27
curve fitting, 325
cut and paste, 25

D

data types, 43
date, 8, 101
deal, 251
Debugger, 174
debugging, 257
degrees to radians, 8
del2, 228
delimiter highlighting, 51
demo, 12
det, 145
determinate repetition, 181
diag, 138
diary, 39
dice, 114
diff, 307
differential equation, 308, 360
differentiation, 360
disp, 52
display
 scale factor, 55
division
 matrix, 156
dot product, 47
dot-transpose operator, 116
double, 44
double precision, 43

doubling time of an investment, 186
drawnow, 224

E

editing a plot, 220
eig, 145
eigenvalue, 148
eigenvector, 149
electrical circuit, 265
element-by-element operation, 46
ellipse, equation of, 234
ellipsis, 42, 48
elseif, 67
empty, 35
empty array, 120
end, 132
eps, 111
error
 fatal, 257
 logical, 258
 rounding, 259
 run-time, 174
 syntax, 257
etime, 61
Euler's method, 308
exclusive OR, 115
exist, 119
exp, 101
explicit list, initializing vectors, 34
expm, 145, 313
exponent, 43
exponential
 matrix, 313
exponential fit, 155
exponential growth, 309
exporting ASCII data
 save, 104
exporting binary data
 save, 105
expression, 42–44, 47
 ans, 47
 display suppressed, 48
 logical, 64, 115
eye, 135

F

factorial, 57
false, 109
fast Fourier transform, 329
feval, 171
Fibonacci numbers, 180
field, 249

fieldnames, 250
figure, 10, 203
figure window, 10, 202
find, 119
findobj, 218
finite difference scheme, 311, 318
fix, 101
fixed point, 43, 54
fliplr, 138
flipud, 138
floating-point, 43, 54
floor, 59, 101
for, 58
 command line version, 60
 index, 59
 iteration count, 59
 more general, 60
 most general form, 139
 nested, 140, 184
 replaced by vectorization, 60
 used for determinate loop, 192
format, 54
formula
 vectorization, 50
Fourier series, 194, 329
fplot, 205
fractal, 235
free fall problem, 265
frequency distribution, 245
full matrix, 157
full, 159
function, 49, 166
 handle for passing as argument, 171
 help, 166
 MATLAB Function Reference, 99
 name resolution, 173
 P-code file, 170
 private, 170
 subfunction, 170
function functions, 172
function M-file, 163
function/command duality, 173
fzero, 303

G

Gauss reduction, 161
Gaussian random number, 293
gca, 215
gcf, 215
gco, 216
get, 216

Gibbs phenomenon, 194
ginput, 203
global, 167
gradient, 211, 212
graph, 9
 easy-to-use commands, 198
graphics
 2-D, 197
 3-D, 206
 3-D rotation, 213
 axis limits, 200
 contour plot, 209
 cropping a surface, 211
 editing, 220
 editing a plot, 220
 line type, 200
 logarithmic plot, 203
 mesh surface, 207
 multiple plots, 199, 202
 plotting symbol, 200
 polar co-ordinates, 204
 rapidly changing functions, 205
 saving, 230
 vector field, 211
grid, 199
gtext, 198, 244
guessing game, 184

H

H1 line, 167
Halley's comet, 234
handle
 function, 171
 graphics object, 214
Handle Graphics objects, 214
Harmonic analysis, 329
help
 for a function, 166
Help Navigator, 99
helpwin, 99
hexadecimal, 54
hidden name, 173, 258
histogram, 245
hold, 199

I

i, 71
identity matrix, 135
if, 63, 64
 command line version, 63
 elseif, 67

nested, 70
 vectorized with logical vector, 121
if-else, 66
 command line, 66
ill conditioning, 155, 161
imag, 72
importing ASCII data
 load, 105
importing binary data
 load, 106
inclusive OR, 115
indeterminate repetition, 181, 184
index of for, 59
Inf, 7
infinity, 7
initialization, colon operator, 35
initialization, explicit list, 34
inline object, 92
input argument, 93
integration, 360
interpolation, 305
interpreter, 28
inv, 145
isempty, 120
isglobal, 167
isinf, 120
isnan, 120
isothermal, 210
iteration count, 59

J

j, 71

K

Knuth, 188

L

least squares, 155
left division, 44, 153, 156
Legendre polynomial, 180
length, 101
Leslie matrix, 147
lighting, 229
limit of sequence, 58
line numbers, 174
line style, 200
linear equations
 over-determined, 154
 residual, 153
 solution, 152
 under-determined, 155
linspace, 36

list
 sorting, 246
load, 105, 106
LODE, 313
log, 101
log10, 102
logarithmic plot, 203
logarithmic spiral, 232
logical error, 258
logical expression, 63, 64, 109, 115
 vector, 110
logical function, 119
logical operator, 68, 115
 precedence, 115
logical vector, 110
 subscripting with, 118
loglog, 204
logspace, 36
lookfor, 167
loop
 determinate, 181
 indeterminate, 184
Lorenz, 314
Lotka-Volterra model, 317
lu, 145

M

M-file, 27
 function, 163
 function passed as argument, 171
magic square, 22, 136
mantissa, 43, 54
marker, 200
Markov chain, 149
MATLAB functions, 92
matrix, 34, 38, 127
 array operation, 138
 creating, 129
 deleting rows and columns, 134
 duplicating rows and columns, 133
 enlarging, 129
 full, 157
 Leslie, 147
 multiplication, 143
 plot, 197
 sparse, 157
 subscript, 129
 transpose, 38, 130
 visualization, 212
matrix division, 156
matrix exponential, 313
matrix exponentiation, 145

matrix multiplication, 143
matrix operation, 46, 143
max, 102, 249
mean, 102
mechanical system, 341
menu, 191
mesh, 208
mesh surface, 207
meshc, 210
meshgrid, 207
Mexican hat, 22
min, 102, 249
mlock, 168
Monte Carlo estimation of π, 294
more, 53, 189
multi-dimensional array, 142
multiple lines in Command Window, 26
multiple plots, 202
multiple plots on same axes, 199
munlock, 168

N

name hiding, 173, 258
NaN, 7
nargin, 169
nargout, 169
nested fors, 140, 184
nested ifs, 70
Newton quotient, 179, 260, 306
Newton's method, 56, 163
Newton's method in general, 299
normally distributed random number, 293
NOT, 115
not-a-number, 7
num2str, 53
number, 43
numerical differentiation, 306
numerical integration, 304
numerical method, 299
 Bisection method, 301
 differential equation, 308
 differentiation, 306
 Euler's method, 308
 integration, 304
 Newton quotient, 306
 Newton's method, 299
 partial differential equation, 318
 predictor-corrector, 312
 Runge-Kutta method, 313
 Simpson's rule, 305

solution of equations, 299
 Trapezoidal rule, 304
numerical solution of differential equations, 308
numerical solution of equations, 299

O

object, 254
 Handle Graphics, 214
object-oriented programming, 254
ones, 135
operand, 44
operation
 array, 46
 element-by-element, 46
 matrix, 143
operator
 arithmetic, 44
 colon, 130
 dot-transpose, 116
 left division, 153
 logical, 68, 115
 precedence, 44, 116
 relational, 64
 transpose, 116
operator precedence, complete list, 116, 385
OR, 115
 exclusive, 115
oscillator, 341
output, 52
 paging, 53
 suppressed with semicolon, 48
output argument, 165
over-determined system, 154

P

P-code file, 170
paging, 189
paging output, 53
partial differential equation, 318
pascal, 136
pause, 214
pcode, 170
persistent variable, 167
pi, 8
plot, 197
 3-D version, 206
 logarithmic scale, 203
 polar co-ordinates, 204
plot edit mode, 220
plot3, 206

plotedit, 220
plotyy, 199
Poisson process, 290
polar, 204
polar co-ordinates, 204
polynomial fitting, 325
pow2, 102
pre-allocation, 239
precedence, 116
 arithmetic operators, 44
 colon operator, 45
 logical operator, 115
 transpose, 46
precedence of operators, complete list, 116, 385
predator-prey model, 317
predictor-corrector method, 312
prime number
 largest known, 188
 test, 187
private function, 170
prod, 102
Profiler, 171
program, 24
projectile problem, 265
projectile trajectory, 188
Property Editor, 206
pseudo-code, 39
pseudo-random number, 102

Q

gr, 145
quad, 306
quadrature, 306
quiver, 211
quote in string, 53

R

radio-active decay, 296
RAM, 28
rand, 102, 113, 135, 285
 seeding, 286
 uniform, 285
randn, 293
random access memory, 28
random number, 37, 113
 generation, 285
 normal, 293
 uniform, 285
random numbers
 counting, 113

random walk, 150, 245, 288
rat, 54
real, 72
realmax, 43, 102
realmin, 43, 102
recall
 smart, 7
recursion, 176
redimensioning, 240
reference, pass by, 169
relational operator, 64
relative accuracy, 43
rem, 102
removing element of vector, 119
removing elements of an array, 37
repmat, 133
reshape, 133
reshaping a matrix, 132
residual, 153
right division, 156
rmfield, 251
rolling dice, 114, 287
root of an equation, 299
roots, 303
rot90, 138
rounding error, 259, 306
run-time error, 174
Runge-Kutta method, 313

S
save, 104, 105
saving a file, 26
saving a graph, 230
scalar, 9
scalar expansion, 47
scalar product, 47
scientific notation, 43
 display, 54
scope, 167
script, 27
 recommended way to run, 28
script file, 27
seeding rand, 286
semi-colon, 7
semilogx, 204
semilogy, 203
sensitivity analysis, 161
separator, 49
series for π, 193
set, 217
Simpson's rule, 305

simulation, 114, 245
 bacteria division, 288
 bacteria growth, 294
 estimation of π, 294
 game of eights, 295
 radio-active decay, 296
 random walk, 288
 rolling a dice, 287
 spinning coins, 286
 traffic flow, 290
single-precision, 44
size, 33
smart indenting, 59
smart recall, 7
sort, 246, 248
sort
 bubble, 246
sound, 183
sparse matrix, 157
sparse, 158, 159, 320
sparse matrix, 320
spinning coins, 286
spiral
 Archimedes, 232
 logarithmic, 232
spring, mass damper model, 341
sprintf, 244
spy, 159, 213
sqrt, 8
stable age distribution, 149
statement, 7, 48, 49
 assignment, 48
 continuation, 48
string, 53
 containing quote, 53
struct, 250
structure, 249
structure plan, 26, 39
subfunction, 170
subplot, 202
subscript, 37, 129
 logical vectors, 118
 vector, 118
subscript out of range, 129
subscripting with logical vectors, 118
sunflower head, 233
surf, 209
surfc, 210
svd, 145
switch, 71
symbolic mathematical, 359
syntax error, 257

T
tab completion, 11
Taylor series
 exponential, 179
terminal speed, 272
text, 199
tic, 61
tiling, 134
title, 199
toc, 61
top down programming, 40
traffic flow, 290
transpose, 37, 38, 130
 complex conjugate, 72
transpose operator, 116
transpose operator, highest
 precedence, 46
Trapezoidal rule, 304
tridiagonal matrix, 320
trigonometry, 360
tril, 138
triu, 138
true, 109
truecolor, 228
truncation error, 325
type, 27

U
under-determined system, 155
uniformly distributed random
 number, 285
update process, 237

V
value, pass by, 169
Van der Pol's equation, 327
varargin, 170, 253
varargout, 253
variable, 7, 28, 31
 global, 167
 persistent, 167
 scope, 167
 termed arrays by MATLAB, 32
vector, 9, 34
 colon operator, 35
 explicit list, 34
 incompatible sizes, 257
 pre-allocation, 239
 redimensioning, 240
 removing element, 119
 scalar product, 47

subscript, 37, 118
transpose, 37
vectorization, 50
 instead of for, 60
view, 213
visualization of matrix, 212

W

while, 185
 command line form, 185
 used for indeterminate loop, 192

vector condition, 185
who, 33
whos, 33
why, 24
workspace, 32
 base, 167
 function, 167
Workspace Browser, 33

X

xlabel, 199

xor, 115

Y

ylabel, 199

Z

zero of a function, 299
zeros, 135
zlabel, 206
zoom, 325